Nanomaterials for Sustainable Hydrogen Production and Storage

Hydrogen is poised to play a major role in the transition toward a net-zero economy. However, the worldwide implementation of hydrogen energy is restricted by several challenges, including those related to practical, easy, safe, and cost-effective storage and production methodologies. Nanomaterials present a promising solution, playing an integral role in overcoming the limitations of hydrogen production and storage. This book explores these innovations, covering a wide spectrum of applications of nanomaterials for sustainable hydrogen production and storage.

- Provides an overview of the hydrogen economy and its role in the transition to a net-zero economy.
- Details various nanomaterials for hydrogen production and storage as well as the modeling and optimization of nanomaterials production.
- Features real-life case studies on innovations in nanomaterials applications for hydrogen storage.
- Discusses both the current status and future prospects.

Aimed at researchers and professionals in chemical, materials, energy, environmental, and related engineering disciplines, this work provides readers with an overview of the latest techniques and materials for the development and advancement of hydrogen energy technologies.

Emerging Materials and Technologies
Series Editor: Boris I. Kharissov

The *Emerging Materials and Technologies* series is devoted to highlighting publications centered on emerging advanced materials and novel technologies. Attention is paid to those newly discovered or applied materials with the potential to solve pressing societal problems and improve quality of life, corresponding to environmental protection, medicine, communications, energy, transportation, advanced manufacturing, and related areas.

The series takes into account that, under present strong demands for energy, material, and cost savings, as well as heavy contamination problems and worldwide pandemic conditions, the area of emerging materials and related scalable technologies is a highly interdisciplinary field, with the need for researchers, professionals, and academics from across the spectrum of engineering and technological disciplines. The main objective of this book series is to attract more attention to these materials and technologies and invite conversation among the international research and design community.

For more information about this series, please visit: www.routledge.com/Emerging-Materials-and-Technologies/book-series/CRCEMT

Nanomaterials for Sustainable Hydrogen Production and Storage

Edited by
Jude A. Okolie
Emmanuel I. Epelle
Alivia Mukherjee
Alaa El Din Mahmoud

CRC Press
Taylor & Francis Group
Boca Raton London New York

CRC Press is an imprint of the
Taylor & Francis Group, an **informa** business

Designed cover image: www.shutterstock.com

First edition published 2024
by CRC Press
2385 NW Executive Center Drive, Suite 320, Boca Raton FL 33431

and by CRC Press
4 Park Square, Milton Park, Abingdon, Oxon, OX14 4RN

CRC Press is an imprint of Taylor & Francis Group, LLC

British Library Cataloguing-in-Publication Data
A catalogue record for this book is available from the British Library

ISBN: 978-1-032-44207-5 (hbk)
ISBN: 978-1-032-44208-2 (pbk)
ISBN: 978-1-003-37100-7 (ebk)

DOI: 10.1201/9781003371007

Typeset in Times
by SPi Technologies India Pvt Ltd (Straive)

Contents

Preface

Hydrogen stands to play a vital role in the transition towards a net-zero economy due to its clean and environmentally friendly nature. However, the worldwide implementation of hydrogen energy is restricted by several challenges, including those related to production and storage. Currently, most of the world's hydrogen demand is met through the steam reforming of natural gas, a process that emits substantial quantities of greenhouse gases. Sustainable hydrogen production, by contrast, requires environmentally friendly resources and must be competitively priced with natural gas-derived hydrogen.

The practicality of hydrogen is further hindered by the lack of practical, easy, safe, and cost-effective storage methodologies. Here, nanomaterials present a promising solution, playing an integral role in overcoming the limitations of hydrogen production and storage. The unique physicochemical properties of nanomaterials enable their applications as catalysts in both areas. This book explores these innovations, covering a wide spectrum of applications of nanomaterials for sustainable hydrogen production and storage. It is organized into nine chapters to provide readers and researchers at all levels with easy access to the information contained within.

In Chapter 1, Rogachuk et al. provide an extensive overview of the hydrogen economy, emphasizing its crucial role in achieving sustainable development goals. They highlight the connection between the hydrogen economy and sustainability objectives, outlining current and future applications in industries such as pharmaceuticals, petrochemicals, aerospace, food, beverages, and transportation.

Chapter 2, by Awotoye et al., explore the progress and advancements of nanomaterials, delineating various types, synthesis methods (including physical, chemical, and biological procedures), and characterization techniques. They discuss the broad applications of nanomaterials across various fields.

Epelle et al., in Chapter 3, discuss major challenges in renewable hydrogen production, including high-temperature requirements for biomass decomposition and water splitting, and the corresponding need for expensive thermal-resistant materials. They provide a comprehensive discussion of the role of nanomaterials in overcoming these challenges, specifically through their application as catalysts to enhance thermochemical process efficiency.

Chapter 4, by Abiodun et al., explore different methods of hydrogen production from biological conversion processes, examining the potential of nanomaterials to enhance these methods. They discuss the advantages, applications, efficacy, and efficiency of biological hydrogen production processes, along with the corresponding nanotechnological devices.

Obande et al., in Chapter 5, explore nanomaterial applications in addressing technological barriers for both electrolytic and photolytic hydrogen production. They discuss the need for novel catalysts, advanced electrolyzer designs, and lower-cost materials, as well as the development of cost-effective and efficient photocatalysts for photolytic hydrogen production. Future research directions in these areas are also presented.

Chapter 6, by Fajimi et al., investigate research works that have carefully modeled the production process of carbon-based nanomaterials (mainly carbon nanotubes) over the last two decades, discussing models such as the simple weight model, kinetic model, and those derived from chemical kinetic models.

In Chapter 7, Yaqub et al. explore several machine learning (ML) tools, including traditional ML and deep learning, used to predict different nanomaterial properties based on experimental data. Tools such as artificial neural networks (ANN), support vector machines (SVM), decision trees, convolutional neural networks (CNN), and deep neural networks (DNN) are explained with examples.

Chapter 8, by Güleç et al., provide a comprehensive understanding of different types of solid-state hydrogen storage nanomaterials, including carbonaceous, metal, and complex hydrides, metal-organic frameworks, and covalent organic frameworks. In addition to discussing hydrogen adsorption capacities, preparation methods, potential developments, and challenges in hydrogen storage for each material are also explored.

Akande et al., in Chapter 9, present the state-of-the-art in using nanomaterials for hydrogen storage. They discuss the types of nanomaterials and the challenges (economic and technical) associated with their use in hydrogen storage, along with an overview of advanced characterization techniques and available computational methods.

This book serves as a valuable resource for anyone interested in the innovative field of hydrogen energy, providing insights, methodologies, and cutting-edge research that can inspire and guide further exploration and development in this critical area of sustainable energy.

Notes on the Editors

Dr. Jude A. Okolie is currently an Assistant Professor at the Gallogly College of Engineering, University of Oklahoma. Dr. Okolie's research focuses on the thermochemical conversion of waste materials to green fuels and the subsequent utilization of hydrochar/biochars for environmental remediation. In addition, his research includes the application of process simulation and artificial intelligence/machine learning to address climate change, environmental pollution, and sustainable agriculture challenges Dr. Okolie has published numerous papers in leading international peer-reviewed journals and books. His papers have been rated as highly cited and downloaded worldwide. He has also been ranked among the world's top 2% most cited scientists by Elsevier BV and Stanford University. He has been granted several prestigious local and international awards, including the George Ira Hanson Energy Award for his work on thermochemical hydrogen production, and the University of Oklahoma Alternative Textbook Grant. Dr. Okolie is the author of the book *Biofuel* and is also an editorial board member of the *Societal Impact* journal.

Dr. Emmanuel I. Epelle is a Chancellor's Fellow at the University of Edinburgh's Institute for Materials and Processes. He received his engineering training from The University of Edinburgh (PhD), Imperial College London (MSc), and the Federal University of Technology, Minna (BEng). Dr. Epelle has published extensively in the fields of process modeling, simulation and optimization of complex chemical processes including biomass-to-biofuels conversion processes. He is a Chartered Engineer, an Associate Fellow of the UK Higher Education Academy and a recipient of several research and innovation awards (such as the Scottish Knowledge Exchange awards, and the CeeD Industry awards). He has widely taught the use of computational methods for Chemical Engineering design and continues to supervise undergraduate and postgraduate degree projects in these subject areas.

Dr. Alivia Mukherjee is a highly accomplished Postdoctoral Research Fellow with a strong background in Chemical and Biological Engineering. She currently serves as a Postdoctoral Research Fellow in the Department of Mechanical Engineering at the University of Alberta. Her research pursuits are supported by renowned funding agencies such as Natural Resources Canada (NRCan), the International Energy Agency (IEA),

Suncor Energy Corp., and the University of Alberta. Her work encompasses critical topics within the realm of sustainability, biomass conversion, and greenhouse gas emissions. Dr. Mukherjee is dedicated to advancing clean energy solutions in Canada through her expertise in low-quality and waste-biomass feedstocks.

She completed her Doctor of Philosophy (PhD) in Chemical and Biological Engineering at the University of Saskatchewan. Her Master of Technology (MTech) in Chemical Engineering, specializing in Petrochemicals and Petroleum Refinery Engineering, was obtained from the University of Calcutta in 2015. Prior to that, she earned her Bachelor of Technology (BTech) in Chemical Engineering from the West Bengal University of Technology in 2013. Dr. Alivia Mukherjee has authored a substantial body of scholarly work, comprising over 20 peer-reviewed journal articles and review papers, showcasing her prolific contributions to the academic and research community.

Alivia Mukherjee has received a plethora of prestigious awards and honors throughout her academic and research career, underscoring her outstanding contributions to the field of chemical engineering and sustainability. Her accolades also extend to her presentation skills, as she clinched the Best Oral Presentation Award at the Chemical Institute of Canada's Energy Symposium in 2022, earning the Energy Division prize at the Canadian Chemical Engineering Conference. She has also been the recipient of various awards at different national and international conferences including the Best Poster Presentation award, and the 3MT Thesis Competition Winner's prize. Furthermore, her global recognition includes the Best Oral Presentation Award at the American Chemical Society Omega's International Conference in NIT-Surat, India. Throughout her career, Dr. Mukherjee has contributed significantly to the field of chemical engineering and sustainability.

Dr. Alaa El Din Mahmoud is an Assistant Professor in the Environmental Sciences Department, Faculty of Science at Alexandria University and the co-founder of Green Technology Lab. Dr. Mahmoud is also the Vice-Chair of the National Committee of the UNESCO-MAB (Man and Biosphere) program in Egypt. He received his bachelor's and master's degrees from Alexandria University in Egypt and his PhD degree from Friedrich-Schiller University Jena in Germany. During his career, he received fellowships from Temple University (USA), University of Texas at Austin (USA), Shanghai University (China), University of Luxembourg, and German Academic Exchange Service DAAD. His research focuses on interdisciplinary environmental issues that are related to sustainability, waste valorization, water/wastewater treatment, circular economy, artificial intelligence/machine learning, biochar, and green nanotechnology. Dr. Mahmoud's teaching activities are oriented to aspects of Environmental Sciences and Environmental Technology.

Dr. Mahmoud is also a "Sustainable Development Ambassador" (top-achiever) for the Egyptian Ministry of Planning and Economic Development as well as a certified trainer. He has been awarded the Alexandria University Award for Academic Encouragement in the Scientific Field (2023) and the Best Young Researcher Award for his scientific contribution from the RanBal Research Institute and Bentham Science (2022). Furthermore, he is a recipient of several awards such as the Fulbright Award 2021, "German Academic Exchange Service: DAAD Award" for conference and lecture travel abroad in two consecutive years (2018 and 2019), the American Chemical Society Award for Environmental Chemistry Division in 2018, Misr El Kheir Foundation award for postgraduates in 2014, and the "Young Scientists' Award" from UNESCO-MAB program in 2013.

Dr. Mahmoud has an outstanding publication track record in peer-reviewed international journals and book chapters as well as an editor for several books. Moreover, Dr. Mahmoud is involved in many collaboratively funded research projects as a principal investigator, a consultant, and a team member. Currently, he is a young editorial member of *Biochar Journal* and *Carbon Research Journal*. Among his current activities, Dr. Mahmoud is a reviewer for several prestigious journals.

Contributors

Okon-Akan Omolabake Abiodun
Department of Wood and Paper
Technology
Federal College of Forestry, Jericho
200284, Nigeria
Forestry Research Institute of Nigeria
Ibadan, Nigeria

Adekunle A. Adeleke
Department of Mechanical Engineering
Nile University of Nigeria
Abuja, Nigeria

Olugbenga Akande
Department of Computer Science and
Electrical Engineering
Handong Global University
Handong, South Korea

Damilola Awotoye
Department of Chemical Engineering
University of Ilorin
Ilorin, Nigeria

Sarah M. Barakat
School of Sustainable Chemical,
Biological and Materials Engineering
University of Oklahoma
Norman, Oklahoma

Emmanuel I. Epelle
School of Engineering, Institute for
Materials and Processes
The University of Edinburgh
Scotland, United Kingdom

Sydney A. Etchieson
School of Sustainable Chemical,
Biological and Materials Engineering
University of Oklahoma
Norman, Oklahoma

Lanrewaju I. Fajimi
Department of Chemical Engineering
University of Johannesburg
Johannesburg, South Africa

Fatih Güleç
School of Chemical and Environmental
Engineering
University of Nottingham
Nottingham, United Kingdom

Peter Ikubanni
Department of Mechanical Engineering
Landmark University
Omu Aran, Nigeria

Ovis D. Irefu
Department of Electrical Engineering
University of North Texas
Denton, Texas

Toheeb Jimoh
Department of Chemical Engineering
University of Ilorin
Nigeria

Edward H. Lester
School of Chemical and Environmental
Engineering
University of Nottingham
Nottingham, United Kingdom

Xin Liu
School of Chemical and Environmental
Engineering
University of Nottingham
Nottingham, United Kingdom

Alaa El Din Mahmoud
Environmental Sciences Department
Alexandria University
Alexandria, Egypt

Alivia Mukherjee
Department of Mechanical Engineering
University of Alberta
Edmonton, Canada

Shahrouz Nayebossadri
School of Metallurgy and Materials
University of Birmingham
Birmingham, United Kingdom

William Oakley
School of Chemical and Environmental
    Engineering
University of Nottingham
Nottingham, United Kingdom

Winifred Obande
Institute for Materials and Processes,
    School of Engineering
The University of Edinburgh
Scotland, United Kingdom

Bilainu O. Oboirien
Department of Chemical Engineering
University of Johannesburg
Johannesburg, South Africa

Chukwuma C. Ogbaga
Department of Biological Sciences
Coal City University
Enugu, Nigeria
and
Departments of Biotechnology
    Microbiology and Biochemistry
Philomath University
Kuje, Abuja, Nigeria

Jude A. Okolie
Engineering Pathways Department
University of Oklahoma
Norman, Oklahoma

Patrick U. Okoye
Instituto de Energías Renovables
Privada Xochicalco S/n Col. Centro.
    Temixco, Morelos, 62580, Mexico

Oluwaseun Iyadunni Oluwasogo
School of Applied Bioscience
Kyungpook National University
Daegu, South Korea

Fredrick O. Omoarukhe
Department of Chemical Engineering
University of Ilorin
Ilorin, Nigeria

Brooke E. Rogachuk
School of Aerospace and Mechanical
    Engineering
University of Oklahoma
Norman, Oklahoma

Emma K. Smith
School of Computer Science
University of Oklahoma
Norman, Oklahoma

Feiran Wang
Centre for Additive Manufacturing
University of Birmingham
Birmingham, United Kingdom

Zainab T. Yaqub
Department of Chemical Engineering
University of Johannesburg
Johannesburg, South Africa

List of Abbreviations

ANN	Artificial neural networks
BG	Bandgap
CaSZ	Calcium-stabilized cubic zirconia
CB	Conduction band
CC	Carbon cloth
CNF	Carbon nanofiber
CNN	Convolutional neural networks
DNN	Deep neural networks
CNT	Carbon nanotube
COF	Covalent organic frameworks
CVD	Chemical vapor deposition
DFT	Density functional theory
GA	General Atomic
GCMC	Grand Canonical Monte Carlo
GN	Graphene nanosheets
GPR	Gaussian process regression
H_2	Hydrogen
H_2O	Water
H_2SO_4	Sulfuric acid
HER	Hydrogen evolution reaction
HyS	Hybrid sulfur cycle
IEA	International Energy Agency
JAEA	Japan Atomic Energy Agency
KOH	Potassium hydroxide
LDH	Layered double hydroxide
ML	Machine learning
MOF	Metal-organic framework
MWCNT	Multi-walled carbon nanotube
MXene	Two-dimensional carbides/nitrides
NC	Nitrogen-doped carbon
NF	Nanofiber
NIR	Near-infrared
NM	Nanomaterial
NMR	Nuclear magnetic resonance
NPs	Nanoparticles
OER	Oxygen Evolution Reaction
Pd	Palladium
PGMs	Platinum-group metals
PO	Partial oxidation
Pt	Platinum
PTG	Power to gas
QDs	Quantum dots

rGO	Reduced graphene oxide
Ru	Ruthenium
S-I	Sulfur iodine cycle
SEM	Scanning electron microscopy
SMR	Steam methane reforming
SVM	Support vector machines
SWCNT	Single-walled carbon nanotube
TCWS	Thermochemical water splitting
TEM	Transmission electron microscopy
TGA	Thermal gravimetric analysis
TMCs	Transition-metal carbides
TMNs	Transition-metal nitrides
TMPs	Transition-metal phosphides
TMSs	Transition-metal sulfides
TRL	Technology readiness level
UV	Ultraviolet
V-Cl	Vanadium chloride cycle
VB	Valence Band
WGS	Water gas shift
XPS	X-ray photoelectron spectroscopy
XRD	X-ray diffraction
YSZ	Yttrium-stabilized cubic zirconia

1 Transition toward a Sustainable Hydrogen Economy
Status and Progress

*Brooke E. Rogachuk, Chukwuma C. Ogbaga,
Sydney A. Etchieson, and Alivia Mukherjee*

1.1 INTRODUCTION

At present the world is facing two interrelated challenges: the depletion of petroleum-based resources, which is aggravated by the elevating energy demand and environmental issues associated with the utilization of petroleum resources [1]. The elevating energy demand, the adverse impact of petroleum resources, environmental pollution, climate change issues, and waste disposal due to anthropogenic activities are major concerns in the present world. To address these challenges, several researchers, government organizations, and policymakers have started exploring alternative energy sources that are sustainable, environmentally benign, and readily available to sustain the increasing population and energy demand.

Some alternative energy production routes that have been discussed recently include renewable energy resources such as solar, wind, geothermal, as well as nuclear energy systems. While renewable energy systems such as wind and solar provide a clean pathway for energy production, they are limited by several challenges. These challenges are due to the intermittency nature of renewable energy systems as they are mostly dependent on climatic conditions [2]. Furthermore, some of these energy systems are characterized by significant land usage and infrastructure requirements. There are also concerns about pollution associated with nuclear wastes. Biomass energy, on the other hand, is an abundantly available and cheap source of energy although there are concerns related to biomass heterogeneity, as well as logistics and transportation issues [3].

Regardless of the sources of energy, the immediate goal is to ensure that there is a sustainable, clear, efficient, and versatile form of energy or energy carriers, that can help alleviate the concerns of overdependency on petroleum resources and environmental challenges associated with its consumption. One of the energy carriers that fits the aforementioned description is hydrogen due to its unique properties [4]. Hydrogen is often seen as an energy carrier and vector due to its diverse applications.

DOI: 10.1201/9781003371007-1

1

The combustion of hydrogen releases water, although there is a tendency to produce smaller amounts of NO_x if it is burned with air at high temperatures.

The concept of a hydrogen economy refers to the use of hydrogen safely, cost-effectively, and sustainably in diverse settings. The transition toward a hydrogen economy represents a pivotal shift in the global energy landscape, as nations strive to reduce their carbon footprint and combat climate change. Hydrogen, a clean, versatile, and abundant energy carrier, has the potential to revolutionize various sectors, including transportation, power generation, and industrial processes [5]. The adoption of hydrogen as a key component in the energy mix requires the development of efficient production, storage, and distribution technologies, as well as policy frameworks and public-private partnerships to promote its widespread use. The hydrogen economy presents a potential threshold to a new era in sustainable energy, with the opportunity not only to reduce greenhouse gas emissions but also to foster economic growth, technological innovation, and energy security for future generations.

A hydrogen economy is a vision for a future energy system where hydrogen serves as a primary energy carrier, displacing the predominant use of fossil fuels, and facilitating the transition to a low-carbon, sustainable society. A description of a hydrogen economy is presented in Figure 1.1. Its potential applications span numerous sectors, such as transportation, power generation, heating, and industrial processes. In leveraging hydrogen's unique properties, the hydrogen economy aims to address global challenges, including climate change, energy security, and air quality, while fostering technological innovation and economic growth.

The aim of the current chapter is to present an overview of a hydrogen economy, explain the interconnection between a hydrogen economy and sustainable development goals as well as the present and futuristic application of hydrogen as a driving force toward promoting sustainable energy transition.

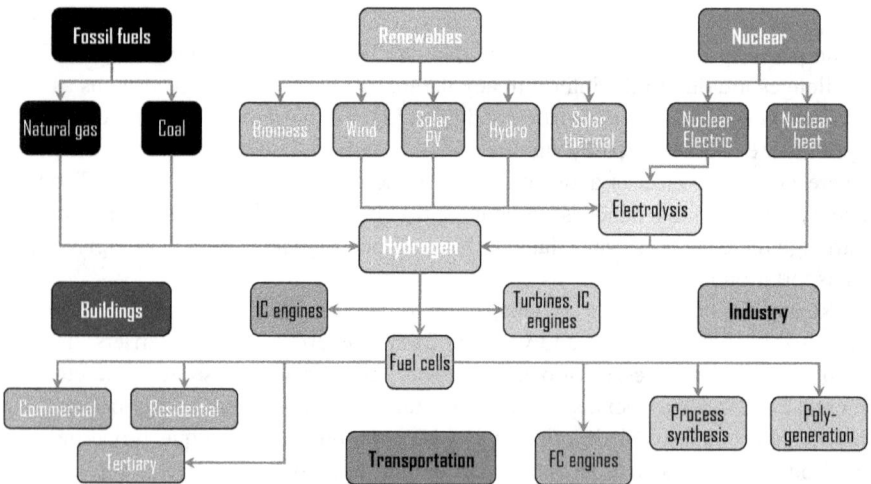

FIGURE 1.1   An overview of a hydrogen economy.

1.2 ADDRESSING SUSTAINABLE DEVELOPMENT GOALS WITH A HYDROGEN ECONOMY

In the next 50 years, the earth's demand for energy will double creating more demand for sustainable options [1]. Also, the issues of climate change, environmental pollution, job security, and clean water have promoted interest in the United Nations sustainable development goals (SDGs). The SDGs have proposed the 2030 Agenda to facilitate the transformation of the world by addressing multiple challenges to ensure well-being, economic prosperity, and environmental protection [6]. Therefore, it is important to systematically analyze the relationship between the hydrogen economy and the SDGs to further understand the position of the hydrogen economy in addressing these challenges. While there are relatively few studies addressing the relationships between the hydrogen economy and the SDGs, some inherent conclusions can be noted. The hydrogen economy has the potential to significantly contribute toward addressing the SDGs, particularly in areas such as clean energy, climate action, and economic growth. Hydrogen has the potential to promote the large-scale integration of renewable energy sources, such as solar and wind, thereby fostering the transition to a low-carbon economy (SDGs 7 and 13). This transition will not only reduce greenhouse gas emissions and combat climate change but also improve air quality and public health in urban areas (SDGs 3 and 11).

Some studies have explored the direct interaction between the 17 SDGs; however, it is still unclear how the hydrogen economy can help foster the SDGs. Pradhan et al. [7] applied the SDG indicator data for 227 countries to study the interactions and trade-offs among the 17 SDGs [7]. It was identified that there is a positive interaction between SDG 1 (no poverty) with almost all other SDGs. However, the extent of such interaction is not clear. Some researchers noted that there are incompatibilities between socio-economic development and environmental sustainability, while also clarifying that some factors such as improved health programs and government investment could play a major role in addressing these incompatibilities [8]. The Weitz et al. [9] study stated that the advancement in SDG target 13.2 (integrating climate change measures into national policies, strategies, and planning) could contribute toward the improvement of clean and affordable energy (SDG 7) [9]. Recently, Akhat et al. [10] demonstrated the economic, social, and geopolitical implications of green hydrogen production from electrolysis powered by renewable energy resources such as wind farms and solar photovoltaic [10]. The research indicates that the manufacture of green hydrogen in South Africa poses substantial risks in a variety of social dimensions, such as child labor, fair remuneration, unemployment, union and negotiation rights, as well as pronounced wage gaps between sexes. However, the risk to these societal elements markedly reduces in other countries when critical equipment is domestically produced rather than imported. This is linked to the increased complexity of the green hydrogen supply chain, as components are sourced from various international locations. Furthermore, the study reaffirms that the interplay between the hydrogen economy and Sustainable Development Goals (SDGs) fluctuates across different countries.

Fostering a hydrogen economy could potentially minimize greenhouse gas (GHG) emissions and potential environmental impacts associated with air pollution [6].

Furthermore, the development and deployment of hydrogen technologies can stimulate innovation, create new employment opportunities, and foster sustainable industrialization in both developed and developing countries (SDGs 8 and 9). By supporting the hydrogen economy, governments, businesses, and research institutions can collectively work toward a more sustainable, equitable, and resilient future for all.

1.3   PRESENT AND FUTURE APPLICATIONS OF HYDROGEN

The global interest in hydrogen stems from its versatility in different applications. One notable use of hydrogen is in the transportation sector as a green fuel. Hydrogen has already proved its ability to be used as a fuel in rocket propulsion systems [11]. Baroutaji et al. [12] comprehensively explored promising applications of hydrogen in the aviation and aerospace industries [12]. They outlined some examples of hydrogen-powered aircraft that use hydrogen as a propellant including the Suntan (USA – 1956), Tupolev Tu-155 (Soviet Union – 1988), CRYOPLANE (Europe – 2000), HyShot (Australia – 2001), NASA X-43 (USA – 2004), Phantom Eye (USA – 2013) [12, 13]. Its high specific energy content makes it a promising fuel for aircraft applications. Rocket fuel tanks are usually charged with liquid hydrogen and liquid oxygen, whose reaction produces a significant propulsion force to drive the rocket upward [12]. Over the years, NASA has consistently used chilled liquid hydrogen as the fuel source for a host of its space vehicles, including but not limited to the Centaur and Apollo [12]. This choice of hydrogen as a fuel has been repeated by numerous rocket engine manufacturers across the globe, including RL10 (Aerojet Rocketdyne – USA), LE-5 (Mitsubishi – Japan), HM7B (Snecma – France), YF-73 (CALT – China), KVD-1 (Russia), and CE-20 (HAL – India) [14].

The aerospace industry is a major contributor to greenhouse gases. An average aircraft flight can contribute 275 grams of CO_2 per passenger [15]. The most common fuel in the commercial airline industry is kerosene. However, if liquid hydrogen were used as a fuel, the overall HC and CO would be close to zero and NO_x would be lowered [16]. The same findings were reported in another study reviewing the effects of kerosene vs. hydrogen as aviation fuel for flights from Toronto to Montreal and Calgary to London [17]. The authors noted that the amount of NO_x, HC and CO emissions for the trips with kerosene-powered aircraft for Calgary were 171.4, 41.9 and 32.2 kg, while for Montreal they were 56.17, 2.43 and 21.9 kg, and for London, they were 251.7, 5.1 and 39.2 kg, respectively. These are significantly lower when compared with liquid hydrogen, demonstrating the potential influence of hydrogen toward achieving a net-zero aviation industry.

Over the last two decades, electric vehicles have steadily grown in popularity. Despite this, the extended time required for charging and the high purchase price of these vehicles remain prohibitive factors, rendering them inaccessible to the broader public. Fuel cells and internal combustion engines both operate in similar manners. Internal combustion engines turn chemical energy into rotational energy that is then used to move the car forward or generate electricity in the vehicle. Fuel cells can do similar things but in a more ecofriendly way. A fuel cell is a device designed to consistently transform chemical energy present in fuel into electrical energy, as long as the fuel and oxidizing agent are supplied. Hydrogen often acts as the fuel source that combines with oxygen to produce electricity, heat, and water as byproducts. Fuel cells' superior traits surpass those of traditional combustion-based technologies currently employed in vital sectors including electronics, residential power, power plants, passenger vehicles, and military applications. Fuel cells operate with greater efficiency than combustion engines, showcasing an electrical energy conversion efficiency of at least 60%, while producing fewer emissions [18]. The oxidation of hydrogen gas in a fuel-cell system produces pure water. Therefore, there are no carbon emissions or air pollutants when hydrogen is used in fuel-cell applications. Hydrogen-powered fuel cells can be used in different settings as indicated in Figure 1.2.

Hydrogen has other applications than as fuel. Hydrogen has vast applications in the pharmaceutical industry. Hydrogen is integral to the manufacture of various raw materials in the pharmaceutical sector, including hydrogen peroxide, sucrose, and hydrochloric acid. For example, hydrogen peroxide is a clear odorless liquid that is widely used as an antimicrobial chemical [19]. Hydrogen peroxide is an effective oxidizing agent compared to other toxic oxidants. It has several applications because it is an environmentally friendly bleach [20]. It has been investigated as a therapeutic medical gas [21]. Hydrogen gas rapidly diffuses tissues and cells without affecting metabolic redox reactions. It regulates gene expression and acts as an anti-inflammatory, anti-allergic, and anti-apoptotic agent [21]. Hydrogen gas can be breathed in through a ventilator to treat oxidative stress without an increase in blood pressure. It can also be used to treat arthritis, diabetes mellitus, and cancer [19].

In the metallurgical industry, hydrogen plays a key role as a metal-reducing agent in extracting metals from their ores as well as for the production of welding-enhanced oxy-hydrogen flames. The oxy-hydrogen flame produced during the high-temperature reaction of hydrogen with oxygen has diverse metallurgical uses including welding and ferrous metal cutting [19].

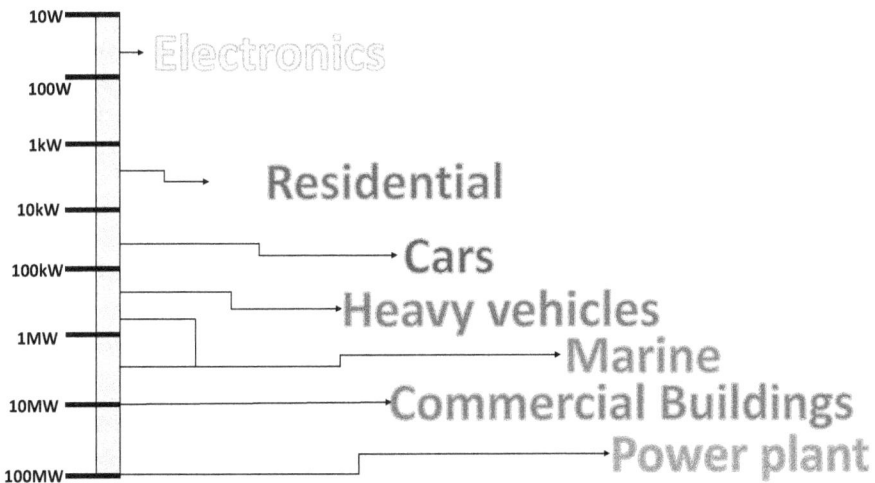

FIGURE 1.2   An overview of different fuel-cell applications.

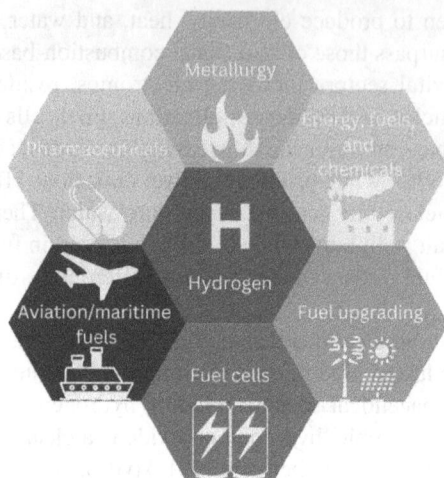

FIGURE 1.3 Pictorial view of different applications of hydrogen.

The isotopes of hydrogen, protium, deuterium, and tritium, are being researched for their different effects in industry. Deuterium has one more neutron than protium and has a smaller molar volume [22]. Deuterium could potentially be used to slow down the metabolism of drugs due to its kinetic isotope effect [19]. However, there is a possibility that deuterium is toxic, but it must be noted that the studies that show this are relatively old [22]. The future possibilities of hydrogen are endless, as presented in Figure 1.3. As more research is conducted, the earth moves closer and closer to a hydrogen economy. Hydrogen and its isotopes continue to prove their value in almost every industry.

1.4 CHALLENGES AND PROSPECTS OF A HYDROGEN ECONOMY

The hydrogen economy faces numerous challenges and opportunities as it endeavors to transform the global energy landscape. One of the primary challenges borders on its efficiency and cost-effectiveness during production, as traditional methods like steam methane reforming and electrolysis can be expensive and emit greenhouse gases. Additionally, safe and efficient storage and transportation of hydrogen pose significant technical and logistical hurdles, as hydrogen is highly flammable and has low energy density by volume. Furthermore, the lack of a widespread hydrogen infrastructure and the need for public and private investments could slow down its adoption.

Liquid hydrogen as a fuel for aircraft can also outperform kerosene in combustion kinetics and maintenance costs. However, hydrogen has low ignition energy and high flame velocity, and it costs more than traditional airline fuels. Once the cost of hydrogen production is reduced and further advancements are made it could be implemented as a main source of airline fuel that is less eco-toxic.

The major challenge of fuel-cell vehicles with hydrogen as the fuel is hydrogen storage. A pressurized tank storage has enough strength for impact resistance and can hold a volume of 186L of hydrogen [11]. However, not enough hydrogen can be stored in gas form. Furthermore, using liquid hydrogen is not currently viable due to the low temperatures needed. Moreover, hydrogen is currently three times more expensive than petroleum. There is also the fear of collisions in hydrogen-fueled vehicles due to the high flammability of hydrogen. In the future, fuel-cell vehicles could replace internal combustion engines after significant research has been performed on the storage of hydrogen. Moreover, advancements in technologies such as fuel cells and electrolyzers are making hydrogen production, storage, and utilization increasingly efficient and economically viable. As the world strives to transition toward a low-carbon future, the hydrogen economy is poised to play a crucial role in facilitating sustainable development and energy security.

1.5 CONCLUSION

The hydrogen economy, a concept referring to the safe, efficient, and sustainable utilization of hydrogen across a variety of contexts, serves as a fundamental shift in how the world approaches energy consumption and production. As global communities strive to lessen their carbon emissions and actively combat the growing threat of climate change, this shift toward a hydrogen-based economy emerges as a vital strategy. This chapter provides an overview of the hydrogen economy, its connection to sustainable development goals (SDGs), and its potential to drastically alter the world's energy landscape. A transition to a hydrogen economy can potentially spur innovation, stimulate job growth, and promote sustainable industrialization. These impacts cater to SDGs 8 and 9, which focus on promoting sustained, inclusive, and sustainable economic growth, full and productive employment, as well as building resilient infrastructure, promoting inclusive and sustainable industrialization, and fostering innovation. Such progress can be achieved in both developed nations, where established industries can leverage these new technologies, and in developing countries, where hydrogen technologies can stimulate growth and modernization. By embracing the hydrogen economy, a collective effort from governments, commercial entities, and research institutions can be harnessed to work toward a future that is not just more sustainable, but also more equitable and resilient. Furthermore, the potential applications of hydrogen in various industries are both vast and forward-thinking. These applications extend beyond mere energy production, touching sectors such as transportation, fuel cells, metallurgy, biomedicine, and pharmaceuticals. Therefore, the shift toward a hydrogen economy offers exciting possibilities for transformative change in a multitude of areas, paving the way for a cleaner and more sustainable future for all.

REFERENCES

[1] F.O. Omoarukhe, E.I. Epelle, C.C. Ogbaga, J.A. Okolie, Stochastic economic evaluation of different production pathways for renewable propylene glycol production via catalytic hydrogenolysis of glycerol, *React Chem Eng.* 8 (2022) 184–198. https://doi.org/10.1039/D2RE00281G

[2] M.H. Alsharif, J. Kim, J.H. Kim, Opportunities and challenges of solar and wind energy in South Korea: A review, *Sustainability* 10 (2018) 1822. https://doi.org/10.3390/SU10061822

[3] A. Tursi, A review on biomass: Importance, chemistry, classification, and conversion, *Biofuel Res J.* 6 (2019) 962–979. https://doi.org/10.18331/BRJ2019.6.2.3

[4] S.A. Sherif, F. Barbir, T.N. Veziroglu, Wind energy and the hydrogen economy—review of the technology, *Solar Energy.* 78 (2005) 647–660. https://doi.org/10.1016/J.SOLENER.2005.01.002

[5] A.E. Sison, S.A. Etchieson, F. Güleç, E.I. Epelle, J.A. Okolie, Process modelling integrated with interpretable machine learning for predicting hydrogen and char yield during chemical looping gasification, *J Clean Prod.* 414 (2023) 137579. https://doi.org/10.1016/J.JCLEPRO.2023.137579

[6] P.M. Falcone, M. Hiete, A. Sapio, Hydrogen economy and sustainable development goals: Review and policy insights, *Curr Opin Green Sustain Chem.* 31 (2021) 100506. https://doi.org/10.1016/J.COGSC.2021.100506

[7] P. Pradhan, L. Costa, D. Rybski, W. Lucht, J.P. Kropp, A systematic study of Sustainable Development Goal (SDG) interactions, *Earths Future.* 5 (2017) 1169–1179. https://doi.org/10.1002/2017EF000632

[8] V. Spaiser, S. Ranganathan, R.B. Swain, D.J.T. Sumpter, The sustainable development oxymoron: Quantifying and modelling the incompatibility of sustainable development goals, *Int J Sustain Dev World Ecol.* 24 (2017) 457–470. https://doi.org/10.1080/13504509.2016.1235624/SUPPL_FILE/TSDW_A_1235624_SM2114.PDF

[9] N. Weitz, H. Carlsen, K. Skånberg, A. Dzebo, SDGs and the environment in the EU: A systems view to improve coherence, (2019). https://policycommons.net/artifacts/1358602/sdgs-and-the-environment-in-the-eu/1971837/ (accessed April 19, 2023).

[10] M.S. Akhtar, H. Khan, J.J. Liu, J. Na, Green hydrogen and sustainable development – A social LCA perspective highlighting social hotspots and geopolitical implications of the future hydrogen economy, *J Clean Prod.* 395 (2023) 136438. https://doi.org/10.1016/J.JCLEPRO.2023.136438

[11] Y. Manoharan, S.E. Hosseini, B. Butler, H. Alzhahrani, B.T.F. Senior, T. Ashuri, J. Krohn, Hydrogen fuel cell vehicles; current status and future prospect, *Appl Sci.* 9 (2019) 2296. https://doi.org/10.3390/APP9112296

[12] A. Baroutaji, T. Wilberforce, M. Ramadan, A.G. Olabi, Comprehensive investigation on hydrogen and fuel cell technology in the aviation and aerospace sectors, *Renew Sust Energy Rev.* 106 (2019) 31–40. https://doi.org/10.1016/J.RSER.2019.02.022

[13] A.G. Galeev, Review of engineering solutions applicable in tests of liquid rocket engines and propulsion systems employing hydrogen as a fuel and relevant safety assurance aspects, *Int J Hydrogen Energy.* 42 (2017) 25037–25047. https://doi.org/10.1016/J.IJHYDENE.2017.06.242

[14] D. Cecere, E. Giacomazzi, A. Ingenito, A review on hydrogen industrial aerospace applications, *Int J Hydrogen Energy.* 39 (2014) 10731–10747. https://doi.org/10.1016/J.IJHYDENE.2014.04.126

[15] N. Jungbluth, C. Meili, Recommendations for calculation of the global warming potential of aviation including the radiative forcing index, *Int J Life Cycle Assess.* 24 (2019) 404–411. https://doi.org/10.1007/S11367-018-1556-3/METRICS

[16] I. Yilmaz, M. Ilbaş, M. Taştan, C. Tarhan, Investigation of hydrogen usage in aviation industry, *Energy Convers Manag.* 63 (2012) 63–69. https://doi.org/10.1016/J.ENCONMAN.2011.12.032

[17] H. Nojoumi, I. Dincer, G.F. Naterer, Greenhouse gas emissions assessment of hydrogen and kerosene-fueled aircraft propulsion, *Int J Hydrogen Energy.* 34 (2009) 1363–1369. https://doi.org/10.1016/J.IJHYDENE.2008.11.017

[18] L. Fan, Z. Tu, S.H. Chan, Recent development of hydrogen and fuel cell technologies: A review, *Energy Reports.* 7 (2021) 8421–8446. https://doi.org/10.1016/J.EGYR.2021.08.003

[19] J.A. Okolie, B.R. Patra, A. Mukherjee, S. Nanda, A.K. Dalai, J.A. Kozinski, Futuristic applications of hydrogen in energy, biorefining, aerospace, pharmaceuticals and metallurgy, *Int J Hydrogen Energy*. 46 (2021) 8885–8905. https://doi.org/10.1016/j.ijhydene. 2021.01.014

[20] S. Sharma, S.K. Ghoshal, Hydrogen the future transportation fuel: From production to applications, *Renew Sust Energy Rev*. 43 (2015) 1151–1158. https://doi.org/10.1016/J. RSER.2014.11.093

[21] S. Ohta, Molecular hydrogen as a preventive and therapeutic medical gas: Initiation, development and potential of hydrogen medicine, *Pharmacol Ther*. 144 (2014) 1–11. https://doi.org/10.1016/J.PHARMTHERA.2014.04.006

[22] T. Pirali, M. Serafini, S. Cargnin, A.A. Genazzani, Applications of Deuterium in Medicinal Chemistry, *J Med Chem*. 62 (2019) 5276–5297. https://doi.org/10.1021/ACS. JMEDCHEM.8B01808/ASSET/IMAGES/LARGE/JM-2018-01808R_0024.JPEG

2 Exploring the Future of Nanomaterials
Insights into Synthesis, Characterization, and Potential Applications

*Damilola Awotoye, Fredrick O. Omoarukhe,
Alaa El Din Mahmoud, Olugbenga Akande,
Adekunle A. Adeleke, Peter Ikubanni, and
Chukwuma C. Ogbaga*

2.1 INTRODUCTION

The significant role of nanotechnology in numerous disciplines has attracted the interest of many researchers globally due to the distinct structures and properties of materials at the nanoscale level. Nanomaterials are materials that have a crystalline or amorphous structure between 1 and 100 nm which is a billionth of a meter (10^{-9} m). Nanomaterials have been highly regarded because of the special properties (physical, chemical, magnetic, etc.) they exhibit [1, 2]. These unique properties make them suitable to be employed in numerous applications that are beneficial to society. Nanotechnology is the science that relates to the synthesis, characterization, and application of nanomaterials. It is an interrelated discipline that features physics, material science, chemistry, engineering, and advanced manufacturing [3].

Based on their properties, nanomaterials can be classified into four types: quantum dots (semiconductor nanocrystals), carbon-related, metal-organic frameworks, and polymer-based nanomaterials [4]. Quantum dots (QDs) are classified as zero-dimensional (0D) nanomaterials because all of their dimensions are less than 100 nm. One-dimensional (1D) nanomaterials like nanowires have one of their dimensions on the nanoscale level and the other two out of the nanoscale level. 2D nanomaterials like nanofilms have two of their dimensions on the nanoscale level, while bulk (3D) nanomaterials like nanocomposites have none of their dimensions on the nanoscale level [5].

The methods for synthesizing nanomaterials are also grouped into top-down and bottom-up methods. The bottom-up approach entails constructing the nanomaterials from clusters of atoms while the top-down method entails reducing or destroying a material to produce nanoparticles [6]. Nanomaterials like carbon nanotubes, graphene,

DOI: 10.1201/9781003371007-2

zinc oxide, copper, and silica nanoparticles are synthesized using different methods like laser ablation, chemical vapor deposition, and many more. These techniques have been divided into three main groups: chemical, physical, and biological methods. Although the physical and chemical approaches are widespread and widely utilized, biological approaches are environmentally sustainable and suitable for the synthesis of nanomaterials to be used in the biomedical industry [7].

It is important to characterize nanomaterials in order to identify their composition and structure and ensure their consistent synthesis. The structural analysis and molecular composition of nanomaterials have been determined using characterization methods that rely on the utilization of instruments like transmission electron microscopy (TEM) [8]. Thermogravimetric analysis (TGA), a mass spectroscopy technique, has also been employed for analyzing the thermal properties of nanomaterials. This is accomplished by monitoring how the materials' masses change with temperature [9, 10]. There are several other techniques used for the characterization of materials including infrared spectroscopy [11], inductively-coupled plasma-mass spectroscopy (ICP-MS) [12], fluorescence spectroscopy [13], surface-enhanced Raman spectroscopy (SERS) [14].

Nanomaterials have been extensively used in different industries because of their special characteristics and small size [15]. For instance, Pan et al. [16] utilized gold nanoparticles to synthesize a keratinocyte growth factor (KGF) nanocomposite which enhanced wound healing and proved to be a promising drug. Other semiconductor nanomaterials have also been proven useful as photothermal agents for treating cancer cells, because of their absorption and photothermal conversion properties [17]. One-dimensional nanoparticles were applied in the renewable energy industries as solar cells due to their structure and physical properties. They are also used in the design of yarn which is currently being considered for the production of fiber-based solar cells [18]. MXenes nanomaterials have also been identified as appropriate materials for heavy metal and organic pollutant removal. They are frequently used in water purification and environmental remediation processes [19].

Due to the great potential of nanomaterials and the rising interest in nanotechnology, it is paramount to consolidate previous research and recent developments occurring in the field of nanotechnology. In 2018, Wang et al. [20] outlined the latest development in the synthesis and use of conjugated polymer nanoparticles. Kumar et al. [21] wrote a review paper on the various methods for characterizing nanomaterials based on their properties. Manzetti and Gabriel [22] presented a review of the progress in carbon nanotubes' (CNTs) synthesis, applications, and modifications. Patel et al. [23] also discussed the recent developments of nanomaterials in prosthodontics. They presented an overview of the effects of various nanomaterials that are utilized in prosthodontics. Saliev [24] conducted a review of the current advances of carbon nanotubes in biomedicine. Their paper encompassed how carbon nanotubes are utilized for treating cancer, in antibacterial therapy, and many more areas.

Numerous reviews are being conducted in the area of nanotechnology to review the current progress being made in the field. This helps researchers to gain new insights and inspiration that promote novel research ideas. Therefore, this chapter presents an overview of the research progress and development of the various types of nanomaterials as well as their synthesis methods and characterization techniques. Furthermore, the applications of nanomaterials in various fields are elucidated.

2.2 TYPES OF NANOMATERIALS

2.2.1 CARBON-BASED NANOMATERIALS

This section provides an overview of carbon-based materials including carbon nanofibers, carbon nanotubes, activated carbon and graphene.

2.2.1.1 Carbon Nanofibers (CNFs)

Carbon nanofibers (CNFs) are organic nanomaterials that have good chemical structure stability, electrical conductivity, and surface area properties. Likened to other carbon-based nanomaterials, CNFs are inexpensive to produce and perform excellently [25]. Electrospinning and chemical vapor deposition (CVD) have been recognized as the most efficient methods for synthesizing CNFs among other techniques like templating, phase separation, and drawing. Depending on the type of synthesis technique and catalytic material, CNFs have different properties and form different shapes (tubular, cylindrical, conical, etc.) with a wide range of diameters [26]. The two major types of carbon nanofibers are herringbone and bamboo fibers. The herringbone fibers have dense graphene-walled conical shapes and large angles between the graphene and fiber layers, while the bamboo fibers have cylindrical shapes and small angles instead [27].

CNFs can also be synthesized to form nanocomposites suitable for the designing of electronic devices and intelligent materials [28]. Qiao et al. [29] recently designed a unique technique for the synthesis of nanofibers using an electrospinning and carbonizing method. This was done to improve their electromagnetism properties and the synthesized CNF nanocomposites were reported to have excellent 3D structures with good electromagnetic properties, making them very suitable for electromagnetic (EM) wave absorption materials.

2.2.1.2 Carbon Nanotubes (CNTs)

Carbon nanotubes are the modified form of carbon nanofibers (CNFs) that are rolled up into perfect cylinders. Based on the number of graphene sheets in their structure, CNTs are divided into single-walled carbon nanotubes (SWCNTs) and multi-walled carbon nanotubes (MWCNTs) groups as shown in Figure 2.1 [30]. The major methods utilized for synthesizing CNTs are laser ablation, carbon arc discharge, and

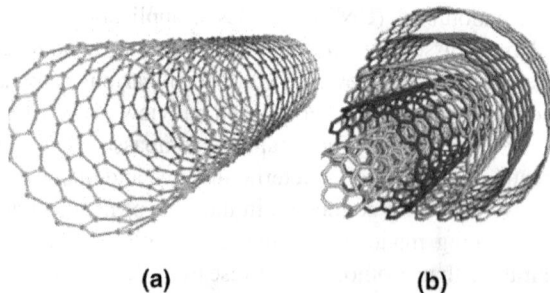

(a)   (b)

FIGURE 2.1   Distinction between (a) SWCNT and (b) MWCNT.

chemical vapor deposition (CVD) which is a thermal synthesis method [31]. In addition, CNTs have excellent electrical and thermal conductivities. They can also act as carriers of different biological molecules which makes them suitable for numerous roles in biomedicine and electrical devices [24]. After ensuring their suitability and biocompatibility with organic materials or metallic nanoparticles, CNTs can be used as biosensors or for effective and targeted drug delivery within cells [32]. They can also be prepared and utilized as acid catalysts in heterogeneous hydrolysis reactions [33]. Due to their excellent electrical properties, SWCNTs have been identified as suitable materials for designing semiconductor electrical devices [34].

2.2.1.3   Activated Carbon

Activated carbon refers to carbonaceous materials produced via thermochemical conversion processes and subsequent activation (physical or chemical). Activated carbon nanoparticles are used in antimicrobial medications to treat infections because of their antimicrobial qualities [35]. Activated carbon can also be combined with carbon nanomaterials like nanofibers or functionalized with metallic ions to improve their properties and performance for various applications including water remediation, catalysis for thermochemical hydrogen production, and electrochemical energy storage [36].

2.2.1.4   Graphene

Graphene materials are 2D crystal allotropes of carbon that appear as sheets of carbon [37, 38]. Graphene-based nanomaterials are nanomaterials with good biocompatibility, optical, mechanical, and electrical properties. They have different physicochemical properties based on the fabrication/synthesis method [39]. They have good mechanical strength, higher surface area, thermal stability, and good electrical properties Graphene-related nanomaterials are currently used as materials for adsorbing different categories of contaminants in water and wastewater. Further details can be found in recent literature [40–43]. Furthermore, these nanomaterials can be used in the creation of flexible semiconductors and supercapacitors [44, 45].

2.2.1.5   Fullerenes

Fullerenes are 3D symmetrical allotropes of carbon that are mainly spherical in shape. They are usually synthesized at high temperatures using synthesis methods like the arc discharge method and their properties are determined by the purity level of the fullerenes [46]. They have pentagonal or hexagonal ring structures made up of interconnected carbon atoms and have numerous types (buckminsterfullerene, bucky clusters, polymers, etc.) depending on their type of structure [47]. Fullerenes have unique properties that make them suitable materials for the designing of sensors, surface-coating materials, solar cells, semiconductors, and electrical devices [48].

2.2.2   Metal-organic Frameworks (MOFs)

These are inorganic nanomaterials that are not formed from carbon but a combination of metal ions and organic ligands. They are also called porous coordination polymers (PCPs). These nanomaterials are synthesized from metals alone or combined with other types of nanomaterials into the suitable nanomaterial range. They are generally porous and consist of zero-valent nanomaterial and metal oxides [49]. They can be used to produce various nanostructured materials after undergoing thermal conversion processes [50].

2.2.3 Metal Nanomaterials (MNPs)

Metal nanomaterials (MNPs) are inorganic nanoparticles synthesized from metals like iron, titanium, zinc, silver, copper, and aluminum, by destruction (top-down) or constructive (bottom-up) methods. Metal NPs usually have an inorganic metal or metal oxide encased in a shell [51]. The major method utilized for synthesizing metal nanoparticles is the chemical reduction method which requires a reducing agent to reduce the divalent and trivalent metal ions [5]. Metal MNPs have distinct surface characteristics and chemical properties. These properties enable metal NPs to be applied in photocatalysis reactions, electronic sensors, and optical coatings, but they can also be harmful to human health depending on their physicochemical properties [52]. The synthesis of metal NPs has become very important as they are widely utilized in various industries [53].

2.2.3.1 Metal-oxide Nanomaterials

To enhance the properties and functionalization of MNPs, metal-oxide nanomaterials are developed from the oxidization of their corresponding metal nanoparticles when it is exposed to oxygen. Metal-oxide nanoparticles have higher reactivity and improved properties than their metal counterparts [54, 55]. Metal and metal-oxide NPs are recognized as suitable materials for hydrogen storage due to their surface area and adsorption properties [56]. Biosynthesis procedures are employed to synthesize both metal and metal-oxide nanoparticles in order to reduce their toxicity and harmful properties. These methods increase the feasibility of using metal-oxide NPs in biomedical applications [57]. Chemical methods like thermal decomposition can also be utilized to synthesize metal-oxide NPs [58].

2.2.4 Semiconductor Nanomaterials

Semiconductor nanomaterials are inorganic nanomaterials that have both metallic and non-metallic properties which make them highly suitable for the formation of semiconductor nanocomposites [59]. Biological methods of synthesis can be utilized for the controlled synthesis of semiconductor nanomaterials. These techniques are utilized to control the crystal structure and size of semiconducting nanomaterials. [60]. Semiconductor NPs decorated with quantum dots have been utilized as efficient photocatalysts that degrade organic pollutants [61].

2.2.5 Polymeric Nanomaterials

Polymeric nanomaterials are commonly referred to as polymer nanoparticles (PNPs). PNPs are organic nanomaterials produced from polymeric materials. Due to the reduced size and surface areas of the polymers, PNPs have improved properties which include higher stability, better intermolecular interactions with surrounding materials, and higher reactivity. [62]. They are widely utilized in the field of nano-medicines, especially for improving the efficiency and delivery of drugs because of their biosafety properties [63].

2.2.5.1 Dendrimers

Dendrimers are 3D synthetic polymeric nanoparticles that have special biological properties. They are highly suitable materials that can be applied in biomedicine and pharmaceutical applications [64]. Their molecular structure exhibits a significant number of branched repeating functional groups. Due to their surface functionality, dendrimers can be utilized as hydrophilic or hydrophobic materials [7].

2.2.5.2 Nanocellulose

Nanocellulose is a nanomaterial obtained by breaking the amorphous structure of cellulose, the most common polysaccharide polymer. It is suitable for creating nano-composites because it is biodegradable and has excellent mechanical qualities [65]. The two types of nanocellulose obtained from cellulose include cellulose nanofibrils and cellulose nanocrystals [66].

2.2.6 CERAMIC NANOMATERIALS

Ceramic nanomaterials are a recent class of non-metallic nanomaterials that are cur-rently being discovered. They are inorganic solid nanoparticles made from materi-als like alumina, titanium dioxide, calcium phosphate, zirconium, and silica [67]. Depending on the method used for their synthesis, ceramic nanomaterials have excel-lent adsorption and photocatalytic properties [68]. They have been found to be excel-lent drug carriers in the biomedical field because of their high thermal resistance and non-reactivity to chemicals [69]. They are also used as bone graft substitutes due to their likeness and biocompatibility with the inorganic constituents of the bone [70]. Ceramic nanofibers, fabricated by electrospinning, are widely utilized in the energy and environmental remediation industries [71].

2.2.7 NANOCOMPOSITES

Nanocomposites are solid nanomaterials that do not have their dimensions confined to the nanoscale level. They are made up of different components and at least one of these components must be a zero-dimensional (0D), one-dimensional (1D), or two-dimensional (2D) nanomaterial [72]. Nanocomposites synthesized from bio-renewable materials like cellulose, starch, or guar gum are being utilized in sen-sors, optical devices, and biomedical applications [73]. Nanocomposites are solid nanomaterials developed by combining a matrix material (e.g., polymer), with a nanofiller/reinforcing material (e.g., nanotubes). This is done to enhance the matrix material's properties and achieve synergy between the combined components [74]. Nanocomposites also have better physicochemical properties than normal compos-ites. Due to their flame retardant, corrosion resistance, and other mechanical proper-ties, nanocomposites have been widely utilized in the space and aerospace industry [75]. Nanocomposites derived from metal oxides are created to improve the photo-catalytic qualities of nanomaterials to be utilized in the semiconductor industry [76].

2.3   NANOMATERIALS SYNTHESIS METHODS

Solid, liquid, and vapor phase techniques have been widely utilized for the production and fabrication of nanomaterials. The numerous methods have been classified into physical, chemical, and biological methods. Each of these methods has its various advantages and drawbacks which are presented in Table 2.1. Past literature on different methods utilized for synthesizing and fabricating nanomaterials are also reviewed and displayed in Table 2.2. As mentioned earlier, the nanomaterials synthesis methods could also be categorized into top-down and bottom-up approaches as presented in Figure 2.2.

2.3.1   PHYSICAL METHODS

Physical methods synthesize nanomaterials without the use of chemicals and, therefore, do not release harmful substances. This leads to the production of high-purity nanomaterials, but they require sophisticated equipment and more time to operate. Physical syntheses are, therefore, labor-intensive and very expensive to run. The physical synthesis of nanomaterials is performed using various approaches including gas-phase deposition, electron beam lithography, powder ball milling, aerosol, and pulsed laser ablation [83, 84].

TABLE 2.1
Advantages and Limitations of Physical, Chemical, and Biological Synthesis Methods

Method of Synthesis	Advantages	Limitations
Physical	– Absence of solvent contamination – Fast and easy synthesis – Synthesized nanoparticles have uniform size distribution – Good purity	– Labor and cost-intensive – Poor agglomeration due to no use of capping agent – High energy requirement – Low stability of nanoparticles – Difficulty in tuning the size and shape of the nanomaterials
Chemical	– Chemical agents help modify desired nanomaterial properties – Low cost – Easy reproducibility – High yield	– Use of harmful chemicals – Low purity of nanomaterials – Long synthesis period – Special conditions required for reaction – Chemical contamination
Biological	– Most environmentally friendly processes – Cost-effective and facile – Highly stable nanoparticles – Non-toxic – Easy availability of biological substrates used for synthesis	– Difficulty in commercialization – Complicated – Difficulty in controlling the characteristics of nanoparticles – Excessive use of microorganisms and plants may lead to an environmental imbalance

TABLE 2.2

Previous Experimental Studies on Nanomaterial Synthesis Methods

Past Literature	Method	Category	Reference
Fabrication of silver (Ag) nanoparticles from silver granules using the gas-phase deposition process	Gas-phase deposition	Physical	[77]
Fabrication of single-walled carbon nanotube (SWCNT) films with the use of dry floating catalysts chemical vapor deposition	Chemical vapor deposition	Chemical	[78]
The use of white rot fungi for the production of silver (Ag) nanoparticles	Biosynthesis with fungi	Biological	[7]
Synthesis of zinc oxide nanoparticles by powder-ball milling method	Powder ball milling	Physical	[79]
Utilization of NaBH$_4$ and poly (vinyl pyrrolidone) (PVP) for the synthesis of Ag NPs	Chemical reduction	Chemical	[80]
Producing zinc oxide (ZnO) NPs with the aid of *Laurus nobilis* L. leaves and zinc salts.	Biosynthesis with plant leaves	Biological	[81]
Experimental fabrication of silver nano disks with electron beam lithography	Electron beam Lithography	Physical	[82]

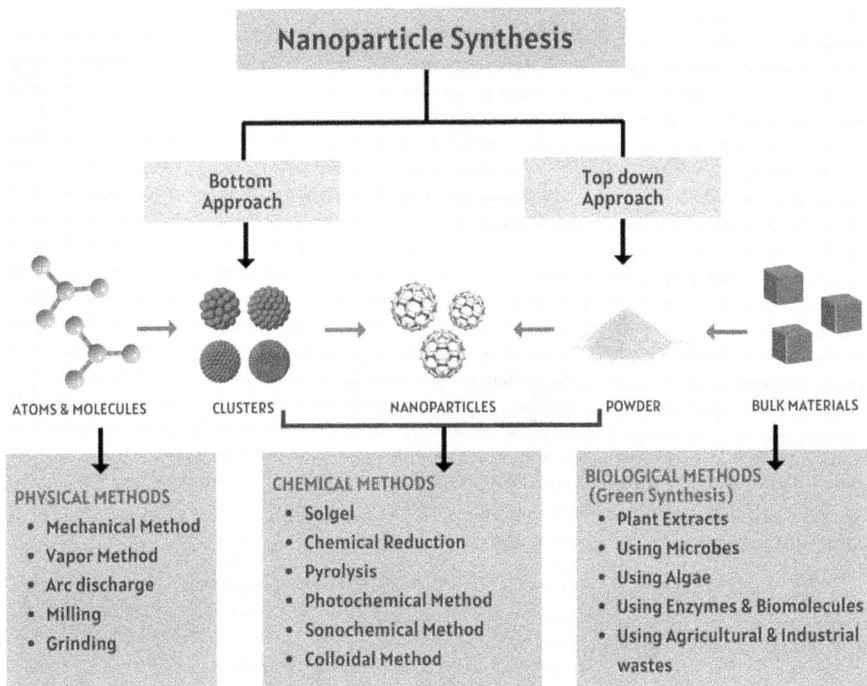

FIGURE 2.2 An overview of nanoparticle synthesis framework.

2.3.1.1 Gas-phase Deposition

This is a physical bottom-up method of synthesizing nanoparticles from atom clusters to their desired size. The initial material is vaporized into an inert gas, after which the gas is cooled, and the vaporized atomic particles are condensed [85]. This method of synthesis is especially utilized for coating different kinds of materials. Gas-phase deposition is also used to synthesize materials with different functionalities as the method allows the mixing of different materials [86]. Thin aluminum oxide (Al_2O_3) films have been applied as coatings for titanium oxide powders using a gas-phase deposition. This was done to reduce the photocatalytic activities of the titanium oxide powder [87].

2.3.1.2 Electron Beam Lithography

This is a photolithography method that utilizes electron beams with short wavelengths for high-resolution patterning of nanomaterials. The technique involves the scanning of a thin organic film layer on the substrate's surface using an electron beam. The unwanted region is then removed using a solvent [88]. This technique is great for the controlled synthesis of nanomaterials with specific structural properties (particle size, spacing, and shape) [89]. The fabrication of nanoarrays on materials to improve their surface-enhanced Raman spectroscopy (SERS) effects has been carried out with the use of electron beam lithography [90].

2.3.1.3 Powder Ball Milling

Powder ball milling is a mechanochemical technique used for synthesizing nanomaterials [91]. It is an appealing technology that is reportedly inexpensive and eco-friendly because it doesn't use solvents [38, 92]. It is a top-down method that involves the breaking down of particles to a particular fineness and particle size [93] (See Figure 2.2). Zinc oxide nanoparticles within the range of 10 nm to 20 nm have been synthesized from commercial zinc oxide powder using the powder ball milling method. The nanoparticles were characterized, and this method has been observed to produce nanoparticles with reduced crystal lattice and size [94].

2.3.1.4 Pulsed Laser Ablation

This is a very simple synthesis technique that is carried out at room temperature. It is an environmentally-friendly and cheap method that is frequently used for the synthesis of carbon-based nanomaterials [95]. This method utilizes a laser pulse or beam to eject particles from a material, for the formation of nanostructures that can be applied in different areas like sensors and biomedicine applications. Pulsed laser ablation is, therefore, a top-down method of synthesis [96]. Pulsed laser ablation has been used for synthesizing cadmium oxide (CdO) NPs from cadmium sheets in water. These nanoparticles have been recognized as suitable materials for producing electrodes that are utilized for energy storage [97].

2.3.2 Chemical Methods

These techniques involve the usage of aqueous solutions/organic solvents as reducing and capping agents for synthesizing nanomaterials from their corresponding metallic salts. Chemical methods are quite flexible as some physiochemical characteristics of the nanomaterials such as dispersion rate, and material structure, can be changed through the chemical agents employed for the synthesis method. Reducing agents are used to reduce the metallic salts to form a colloidal dispersion in the solvent while the capping agents are used to cluster the reduced atoms. There are numerous chemical methods including coprecipitation, microemulsion, hydrothermal, thermal decomposition, electrochemical deposition, and chemical reduction [98–100].

2.3.2.1   Coprecipitation

This is a wet-synthesis technique that involves four stages including nucleation, growth, coarsening, and stabilization stages. The method employs the use of precursors to saturate the aqueous solution and precipitate the nanomaterials [101]. Coprecipitation has been widely applied as a synthesis method that produces metal-based nanoparticles. These nanoparticles have been discovered as suitable nano-adsorbent materials that can be employed for the elimination of heavy metals such as phosphate [102]. A single-step coprecipitation technique is also widely utilized for synthesizing metal-oxide nanoparticles such as zinc oxide, iron oxide, magnesium oxide, among others. [103, 104].

2.3.2.2   Microemulsion

Microemulsion is a thermodynamically stable method of synthesis that involves the formation of an interface layer between a polar and non-polar phase. This interface layer is utilized as a nanoreactor [105]. A chemical reaction occurs in this phase that converts the precursors into nanomaterials. This technique is very versatile and widely utilized for the controlled synthesis of nanoparticles to a desired shape and size [106].

2.3.2.3   Chemical Reduction

This synthesis technique is a frequently used chemical method for synthesizing nanoparticles. The simple technique involves the use of reducing agents (to provide electrons for reducing the substrate) and stabilizing agents (to control the synthesis of the nanomaterial and prevent aggregation) [107]. They are also utilized as a post-synthesis treatment method that will help enhance the functionality of nanomaterials [108].

2.3.2.4   Hydrothermal Synthesis

Hydrothermal synthesis is a simple and widely utilized method for synthesizing nanomaterials from hydrolysis reactions within a wide range of high temperatures. This method employs the use of aqueous solutions and controlled thermodynamic process variables for crystallizing materials [109]. It is ideal for producing big, high-quality crystals and materials that are unstable close to their melting points. The process is utilized for synthesizing inorganic nanomaterials [110]. Highly stable quantum dots have been synthesized from waste (peanut shells and tea leaves) using this method [111].

2.3.3   Biological Methods

Biological methods of synthesizing nanomaterials are also referred to as green synthesis or biosynthesis methods. They are a less expensive, environmentally friendly, and biologically safe option for synthesizing nanomaterials [112]. Biological substrates like fungi, yeast, algae, microorganisms, plant extracts, bacteria, and enzymes, are employed during this synthesis method. These naturally occurring substrates act as stabilizers and reducing agents and, therefore, eliminate the need for chemical agents [113–115]. The absence of toxic chemicals makes biological syntheses not harmful as they do not release or require the use of toxic materials [116]. Biological syntheses yield nanomaterials with special characteristics like better yield, solubility, and stability. These nanomaterials are extensively useful in biological materials because of their biocompatibility and purity. Nanomaterials obtained from synthesis methods with the use of algae, bacteria, plants, yeast, and fungi have been widely applied in the biomedical industry [117, 118].

2.3.3.1 Fungi

Fungi have been identified as a prominent biological substrate for biosynthesis because of their metal bioaccumulation properties and the special ability of fungi to secrete protein, which increases the production of nanoparticles [119]. Fungi have the ability to produce a large number of nanoparticles. The secretion abilities of fungi act as stabilizers and reducing agents and enable the extracellular synthesis of nanoparticles [120]. Fungi have been used to synthesize magnetic nanoparticles including magnetite and iron NPs [121].

2.3.3.2 Yeasts and Plants

Yeasts are capable of absorbing a significant number of dangerous metals from their surroundings due to their large surface area and high tolerance to toxic substances. They are currently utilized for synthesizing metal-based nanomaterials [122]. Plants have been acknowledged as the most promising materials for the biological synthesis of plants due to their availability. Their various parts (leaves, roots, fruits, etc.) have been used to synthesize nanomaterials. For instance, the leaf and root extracted from ginseng have been utilized to synthesize gold (Au) and silver (Ag) nanoparticles [123, 124].

2.4 CHARACTERIZATION OF NANOMATERIALS

The characterization of nanomaterials is vital as it deals with the investigation and study of their structures, chemical constituents, and various properties. This is necessary for the identification, quantification, and reproducible synthesizing of nanomaterials [21]. Characterization techniques are used to identify the size distribution, concentration, reaction rate, diffusivity, solubility, size, and other properties of nanomaterials. This section reviews six different techniques that are currently utilized for the characterization of nanomaterials.

2.4.1 Raman Scattering (RS) Technique

Raman scattering (RS) is a vibrational spectroscopy that is used for detecting the chemical structure and molecular characteristics of nanomaterials. It uses monochromatic light to excite the photons of the nanomaterials and measures the frequency change that occurs after the molecular vibration [125]. Physicochemical properties like the thermal conductivity and thermal resistance of nanomaterials like graphene can be detected using this technique [126]. Due to the low efficiency of Raman spectroscopy in detecting single molecules, nanostructures like plasmonic-magnetic silica nanotubes [127] have been used to enhance their sensitivity leading to a surface-sensitive technique known as surface-enhanced Raman spectroscopy (SERS) [128]. SERS can detect the individual molecules in a nanomaterial. Although RS is suitable for characterizing carbon-based nanoparticles, it loses accuracy at lower concentrations and in complex matrices, which typically contain more carbon [129]. Raman spectroscopy has been utilized to characterize and analyze the number of layers, strain, temperature dependence and doping concentration of graphene [130]. Raman scattering technique is also used to detect the crystalline phases of metal-oxide nanoparticles. Microextraction and Raman spectroscopy were utilized to detect the concentration of titanium oxide (TiO_2) nanoparticles in milk powder [131].

2.4.2   X-ray Diffraction Technique

X-ray diffraction (XRD) is a structural analytical technique used to identify the crystalline formation and molecular structure of materials with the use of an instrument called the X-ray diffractometer. It is a well-established spectroscopic technique that uses an X-ray beam to cause interference on a rotating sample (nanomaterial), which must be in a uniform powdery state [129]. The XRD technique can be utilized to detect characteristics of nanomaterials that are dependent on their shape. It was used to study the hexagonal nanostructures of zinc oxide [132] and has also been used to analyze the size of crystals and phase purity of materials that are employed as cathodes in lithium-ion batteries [133]. Metal-oxide nanoparticles applied in the cosmetic industry as sunscreen filters were characterized using XRD and the crystallite properties and particle size of the nanomaterials were easily obtained directly from the sunscreen sample [134].

2.4.3   Scanning Electron Microscopy Technique

Scanning electron microscopy (SEM) is a popular characterization technique for analyzing and assessing the topography, physical properties, and morphology of nanomaterials [135, 136]. It is particularly useful for observing the dimension and form of nanomaterials and was initially developed to adjunct the properties of both the optical and transmission microscope [133]. The electron microscope detects the scattered electrons on the surface of electrically conductive materials by passing an electron beam over the material's surface [137]. The SEM technique has been utilized to analyze the surface structure of silver nanoparticles which were used as coatings for pellets [138]. A very important parameter for nanomaterial characterization is the size which determines if the nanomaterial falls in the nanoscale level. It is also very important to develop nanomaterials with uniform size distribution, and SEM has been utilized for the detection of commercial zinc oxide (ZnO) NPs' particle deposition and size [139].

2.4.4   UV-VISIBLE SPECTROSCOPY

This is an absorption-based characterization technique that is employed for a wide-spread range of nanomaterials depending on their capacity to absorb ultraviolet light rays [140]. It is an easy and low-cost technique that measures the change in absorbance of the nanomaterials at given wavelengths in relation to their concentrations [141]. UV-visible spectroscopy (UV-vis) can be utilized to characterize nanomaterials and obtain information on their size, agglomeration, shape, and concentration [142]. The UV-visible spectrophotometer has been utilized to monitor the arrangement and absorption of silver NPs in a colloidal solution [143]. It is used to observe the optical properties of polymer nanocomposites and semiconductor nanocrystals so they can be utilized in bioimaging and biosensing devices [144]. In addition, UV-vis has been used to characterize metal nanoparticles fabricated into polymer microgels to evaluate their properties for photonic applications [145].

2.4.5   PARTICLE SIZE ANALYZER

Particle size analysis is an analytical technique that uses an instrument to assess the size and size distribution of nanomaterials. The instrument used is known as a particle size analyzer. Mirzaei et al. [80] made use of a particle size analyzer to measure the particle size distribution of silver (Ag) NPs. This was done to observe the sizes of the particles during the size-controlled synthesis of silver nanoparticles. Particle size analyzers have also been utilized to characterize catechins-Zn complex-loaded nanoparticles. The analyzer determined their particle size distribution, zeta potential, and particle size [146].

2.4.6   FOURIER TRANSFORM INFRARED SPECTROSCOPY (FTIR)

Fourier transform infrared spectroscopy (FTIR) is a widely known method of infrared spectroscopy. It is a vibrational spectroscopy that identifies the surface composition and structure of non-aqueous organic and inorganic nanomaterials [147]. It uses an infrared ray to excite a particle and then measures the molecular vibration and rotation at a wavelength. Unlike other infrared spectroscopy methods, FTIR measures the molecular vibration of particles within a broad range of frequencies [148] as well as detecting the existence of functional groups and purity of nanomaterials [81].

2.5   FUTURISTIC APPLICATIONS OF NANOMATERIALS

It is undeniable that nanotechnology has a lot of relevance in numerous applications because of the distinct characteristics of nanomaterials. This section reviews the current applications of nanomaterials in hydrogen production, pharmaceutical applications, catalysis, wastewater treatment, hydrogen storage, biomedical applications, and the production of green chemicals.

2.5.1   HYDROGEN PRODUCTION AND STORAGE

Hydrogen, a very important and sustainable energy fuel, can be produced by different methods including electrolysis, dark fermentation, photocatalysis, thermochemical water splitting, and many more [149]. Nanotechnology has been applied in diverse ways to aid and improve the production of hydrogen. Inorganic nanoparticles such as titanium oxide, silver, nickel, and iron, have a higher surface area that increases their reactivity and makes them suitable catalysts [6]. Inorganic nanoparticles have, therefore, been employed as catalysts to aid hydrogen production. They are used as suitable catalysts in the dark fermentation process, a method of producing biohydrogen, and have been discovered to be very useful in improving the yield of biohydrogen and the efficiency of the process [150]. Metal and metal-oxide nanoparticles have also been used as catalysts to aid the anaerobic digestion of microalgal biomass and they have shown an enhancement effect on biogas productivity [151]. Nanocatalysts such as graphene are used in photocatalytic reactions such as water splitting to produce hydrogen in a clean and sustainable manner [152]. Nickel oxide and hematite nanoparticles have been utilized to increase the yield of hydrogen that is being recovered from dairy waste [153]. Novel 2D nanoparticles like nanorods have also been fabricated to improve the hydrogen evolution reaction during photocatalytic water splitting, which is another method for the production of hydrogen [154].

There are many challenges when it comes to storing hydrogen because of its low critical temperature and low energy content per volume. Chemical hydrides, adsorbents, metal hydrides, and physical tanks are all different mediums that are used to store hydrogen [155]. The United States Department of Energy (DOE) has set certain conditions that a material must satisfy before it can be utilized to store hydrogen [156]. No material has been recognized to satisfy all these conditions and this has led researchers to utilize nanotechnology to increase the hydrogen storage capacity of various materials [157]. Carbon nanotubes, fullerenes, metal-CNT hybrids, and metal-organic frameworks have been utilized through the physisorption process to improve hydrogen storage. This process allows the molecules of hydrogen to weakly bind on the surface of this nanomaterials [158]. Also, nanomaterials formed from three metals (trimetallic) have been discovered as the most suitable materials for hydrogen storage [159].

2.5.2 WASTEWATER TREATMENT

The importance of wastewater treatment has led to the discovery of numerous technologies that can be used to remove undesired contaminants from water for its use, reuse, or disposal [160–162]. Adsorption has been identified as the most effective technique for treating water among these technologies [163–166]. Due to the special properties of nanomaterials like their high porosity, high surface area, high absorptivity, chemical stability, improved reactivity, reusability, and low energy consumption, they are more appropriate for wastewater treatment [167]. Nanomaterials such as nanocomposite membranes and nanofibers have also been utilized to degrade organic/ inorganic pollutants for their easy removal from contaminated water [168]. Carbon nanomaterials like CNTs and graphene are widely utilized as nano-adsorbents to eliminate both organic and inorganic contaminants from wastewater [169, 170]. Ceramic nanomaterials are also functionalized as adsorbents to adsorb heavy metals such as cadmium. This functional nanomaterial can be used in efficiently treating industrial wastewater streams [171].

2.5.3 PHARMACEUTICAL APPLICATIONS

Nanotechnology has been utilized in the pharmaceutical industry to develop materials that will improve drug delivery systems and create efficient diagnostic tools. These nanomaterials have also been widely used for improving the pharmacological efficacy of drugs. Polymer nanoparticles (PNPs) have been applied as shields for drugs to prevent their deterioration before they get to the targeted site. These PNPs also help protect the cells and tissues from the drug's harmful side effects while also improving their intracellular penetration across cell membranes [172]. Silver nanoparticles are applied in improving cancer treatment due to their anticancer property [173] and by using them in enhanced and targeted drug delivery systems [174]. Gold nanoparticles have also been applied in clinical analysis devices to aid the early diagnosis of diseases and illnesses. They are also utilized to directly target tumor cells for effective treatment and enhance drug delivery systems too [175].

2.5.4 BIOMEDICAL APPLICATIONS

Nanotechnology has facilitated the development of biomedicine due to the functionality and versatility of nanomaterials. In the field of prosthodontics, researchers have observed huge improvements in materials after they have been adjusted into the nanometer range. Wang et al. [176] reported that altering titanium alloy for dental implants to the nanoscale level improved the biological integration of the surface of the implant with the surrounding tissues. Fullerenes have been discovered to be excellent materials suitable for accelerating wound healing. This is due to their anti-oxidative and anti-inflammatory properties which help neutralize pathological mechanisms that hinder the wound healing process [177]. Silver nanoparticles also have excellent antibacterial properties, and this makes them good antimicrobial agents. They also ensure microorganisms do not build tolerance to antibiotics because they are highly effective and reduce the dosage frequency [178]. Nanomaterials are also utilized to improve medical diagnostics. Quantum dots with special optoelectrical properties are combined with magnetic resonance imaging (MRI) to obtain higher contrast images of tumor/cancer sites at lower costs [179]. Carbon nanotubes have also been recognized as exceptional nanoprobes that can be applied in biosensors [180].

2.5.5 CATALYSIS

Nanomaterials have high specific surface area and small pore openings due to their small size. They are, therefore, suitable materials for preparing effective catalysts [181]. Liu et al. [182] prepared carbon nanotubes as solid magnetic arrays due to their stability and magnetic properties. They reported that the magnetic carbon nanotube array performed excellently as a catalyst for the hydrolysis of polysaccharides. Ultrafine metal nanoparticles have also been immobilized to produce high-surface-area nanomaterials that can be applied as nanocatalysts. This is due to their clean surface that improves catalytic performance [183]. Nanocatalysts have also been discovered to be reusable. To catalyze the conversion of 4-nitrophenol into

4-aminophenol, copper nanoparticles were used as catalysts. These copper nanoparticles did not lose any of their catalytic properties despite their repeated reuse [184]. Nanomaterials have also been utilized to improve the catalytic performance of catalysts. Graphene was used to change the catalytic properties of metal catalysts by synthesizing graphene-encapsulated metal nanoparticles [185].

2.5.6 GREEN CHEMICAL PRODUCTION

Nanotechnology is applied in the chemical industry to design sustainable products and technologies that reduce and do away with the usage of dangerous chemicals. This has also advanced numerous applications by reducing toxicity, minimizing cost, and improving the efficiency of chemical processes [186, 187]. Heterogenous catalysts like strong acids and bases have been replaced by nanocatalysts to improve the mechanism of homogenous reactions and protect sensitive functional groups involved, in a clean and sustainable way. Ruthenium hydroxide catalysts have been applied to overcome the limitations of the nitrile hydration reactions, which are used to produce amides [188]. Nanomaterials like zinc oxide are also applied in the design of chemical sensing devices to efficiently improve their detection [189, 190]. Additionally, they are used to prepare materials with a high surface area that are capable of absorbing toxic substances and accelerating the absorption process [191].

2.6 CONCLUSIONS

The application of nanotechnology will be vital for materials in the future. This is because nanoparticles are useful in a wide range of applications because of their distinct properties. Nanotechnology has the potential to enhance the functional characteristics of various types of materials. A promising method for synthesizing nanomaterials is the green synthesis/biosynthesis method. The advantages and drawbacks of the various synthesis techniques were considered in this chapter. Proper characterization of nanomaterials is crucial for understanding their functionality, ensuring their reproducible synthesis, and enabling their effective utilization in a broad range of fields. The benefits and contributions of nanotechnology to human life are amply demonstrated by the usage of nanomaterials in different biomedical and pharmaceutical applications. New avenues for the synthesis, evaluation, and application of various nanomaterials, are greatly aided by this review.

REFERENCES

1. Singh, N.A., Nanotechnology definitions, research, industry and property rights, in *Nanoscience in Food and Agriculture 1*. 2016, Springer. p. 43–64.
2. Mahmoud, A.E.D., Nanomaterials: Green synthesis for water applications, in *Handbook of Nanomaterials and Nanocomposites for Energy and Environmental Applications*, O.V. Kharissova, L.M.T. Martínez, and B.I. Kharisov, Editors. 2020, Springer International Publishing: Cham. p. 1–21.
3. Bhushan, B., Introduction to nanotechnology, in *Springer Handbook of Nanotechnology*. 2017, Springer. p. 1–19.

4. Zhang, L., Applications, challenges and development of nanomaterials and nanotechnology. *Journal of the Chemical Society of Pakistan*, 2020. **42**(5).

5. Kolahalam, L.A., et al., Review on nanomaterials: Synthesis and applications. *Materials Today: Proceedings*, 2019. **18**: p. 2182–2190.

6. Ealia, S.A.M. and M. Saravanakumar. A review on the classification, characterisation, synthesis of nanoparticles and their application. in IOP conference series: materials science and engineering. 2017. IOP Publishing.

7. Gudikandula, K. and S. Charya Maringanti, Synthesis of silver nanoparticles by chemical and biological methods and their antimicrobial properties. *Journal of Experimental Nanoscience*, 2016. **11**(9): p. 714–721.

8. Jo, J., et al., Real-time characterization using in situ RHEED transmission mode and TEM for investigation of the growth behaviour of nanomaterials. *Scientific Reports*, 2018. **8**(1): p. 1–10.

9. Loganathan, S., et al., Chapter 4 – Thermogravimetric analysis for characterization of nanomaterials, in *Thermal and Rheological Measurement Techniques for Nanomaterials Characterization*, S. Thomas, et al., Editors. 2017, Elsevier. p. 67–108.

10. Bertolini, T., D. Fungaro, and A. Mahmoud, The influence of separately and combined bentonite and kaolinite as binders for pelletization of NaA zeolite from coal fly ash. *Cerâmica*, 2022. **68**: p. 375–384.

11. Kiefer, J., et al., Characterization of nanoparticles by solvent infrared spectroscopy. *Analytical Chemistry*, 2015. **87**(24): p. 12313–12317.

12. Montaño, M.D., et al., Single particle ICP-MS: Advances toward routine analysis of nanomaterials. *Analytical and Bioanalytical Chemistry*, 2016. **408**(19): p. 5053–5074.

13. Zhou, J., et al., Single-particle spectroscopy for functional nanomaterials. *Nature*, 2020. **579**(7797): p. 41–50.

14. Wang, X. and L. Guo, SERS activity of semiconductors: crystalline and amorphous nanomaterials. *Angewandte Chemie International Edition*, 2020. **59**(11): p. 4231–4239.

15. Mahmoud, A.E.D. and M. Fawzy, Nanosensors and nanobiosensors for monitoring the environmental pollutants, in *Waste Recycling Technologies for Nanomaterials Manufacturing*, A.S.H. Makhlouf and G.A.M. Ali, Editors. 2021, Springer International Publishing: Cham. p. 229–246.

16. Pan, A., et al., Topical application of keratinocyte growth factor conjugated gold nanoparticles accelerate wound healing. *Nanomedicine: Nanotechnology, Biology and Medicine*, 2018. **14**(5): p. 1619–1628.

17. Huang, X., et al., Design and functionalization of the NIR-responsive photothermal semiconductor nanomaterials for cancer theranostics. *Accounts of Chemical Research*, 2017. **50**(10): p. 2529–2538.

18. Sun, H., et al., Recent progress in solar cells based on one-dimensional nanomaterials. *Energy & Environmental Science*, 2015. **8**(4): p. 1139–1159.

19. Yu, S., et al., MXenes as emerging nanomaterials in water purification and environmental remediation. *Science of the Total Environment*, 2022. **811**: p. 152280.

20. Wang, Y., L. Feng, and S. Wang, Conjugated polymer nanoparticles for imaging, cell activity regulation, and therapy. *Advanced Functional Materials*, 2019. **29**(5): p. 1806818.

21. Kumar, P.S., K.G. Pavithra, and M. Naushad, Characterization techniques for nanomaterials, in *Nanomaterials for Solar Cell Applications*. 2019, Elsevier. p. 97–124.

22. Manzetti, S. and J.-C.P. Gabriel, Methods for dispersing carbon nanotubes for nanotechnology applications: Liquid nanocrystals, suspensions, polyelectrolytes, colloids and organization control. *International Nano Letters*, 2019. **9**(1): p. 31–49.

23. Patel, R.M., et al., Applications of nanotechnology in prosthodontics. *Journal of Evolution of Medical and Dental Sciences*, 2020. **9**(47): p. 3566–3572.

24. Saliev, T., The advances in biomedical applications of carbon nanotubes. *C*, 2019. **5**(2): p. 29.
25. Zhang, B., et al., Recent advances in electrospun carbon nanofibers and their application in electrochemical energy storage. *Progress in Materials Science*, 2016. **76**: p. 319–380.
26. Mohamed, A., Synthesis, characterization, and applications carbon nanofibers, in *Carbon-based Nanofillers and Their Rubber Nanocomposites*. 2019, Elsevier. p. 243–257.
27. Klein, K.L., et al., Surface characterization and functionalization of carbon nanofibers. *Journal of Applied Physics*, 2008. **103**(6): p. 3.
28. Wu, H., et al., Negative permittivity behavior in flexible carbon nanofibers – Polydimethylsiloxane films. *Engineered Science*, 2022. **17**: p. 113–120.
29. Qiao, J., et al., Design and synthesis of TiO2/Co/carbon nanofibers with tunable and efficient electromagnetic absorption. *Chemical Engineering Journal*, 2020. **380**: p. 122591.
30. Shoukat, R. and M.I. Khan, Carbon nanotubes: A review on properties, synthesis methods and applications in micro and nanotechnology. *Microsystem Technologies*, 2021. **27**(12): p. 4183–4192.
31. O'Connell, M.J., *Carbon Nanotubes: Properties and Applications*. 2018: CRC Press.
32. Simon, J., E. Flahaut, and M. Golzio, Overview of carbon nanotubes for biomedical applications. *Materials*, 2019. **12**(4): p. 624.
33. Mallakpour, S. and S. Soltanian, Surface functionalization of carbon nanotubes: Fabrication and applications. *RSC Advances*, 2016. **6**(111): p. 109916–109935.
34. Rao, R., et al., Carbon nanotubes and related nanomaterials: Critical advances and challenges for synthesis toward mainstream commercial applications. *ACS Nano*, 2018. **12**(12): p. 11756–11784.
35. Lakshmi, S., P.K. Avti, and G. Hegde, Activated carbon nanoparticles from biowaste as new generation antimicrobial agents: A review. *Nano-Structures & Nano-Objects*, 2018. **16**: p. 306–321.
36. Cheng, F., et al., Boosting the supercapacitor performances of activated carbon with carbon nanomaterials. *Journal of Power Sources*, 2020. **450**: p. 227678.
37. Tiwari, S.K., et al., Graphene research and their outputs: Status and prospect. *Journal of Science: Advanced Materials and Devices*, 2020. **5**(1): p. 10–29.
38. Mahmoud, A.E.D., A. Stolle, and M. Stelter, Sustainable synthesis of high-surface-area Graphite oxide via dry ball milling. *ACS Sustainable Chemistry & Engineering*, 2018. **6**(5): p. 6358–6369.
39. Liao, C., Y. Li, and S.C. Tjong, Graphene nanomaterials: Synthesis, biocompatibility, and cytotoxicity. *International Journal of Molecular Sciences*, 2018. **19**(11): p. 3564.
40. Mahmoud, A.E.D., Graphene-based nanomaterials for the removal of organic pollutants: Insights into linear versus nonlinear mathematical models. *Journal of Environmental Management*, 2020. **270**: p. 110911.
41. Mahmoud, A.E.D., et al., Biogenic synthesis of reduced graphene oxide from Ziziphus spina-christi (Christ's thorn jujube) extracts for catalytic, antimicrobial, and antioxidant potentialities. *Environmental Science and Pollution Research*, 2022. **29**: p. 89772–89787.
42. Mahmoud, A.E.D., et al., Facile synthesis of reduced graphene oxide by Tecoma stans extracts for efficient removal of Ni (II) from water: batch experiments and response surface methodology. *Sustainable Environment Research*, 2022. **32**(1): p. 22.
43. Mahmoud, A.E.D., M. Franke, and P. Braeutigam, Experimental and modeling of fixed-bed column study for phenolic compounds removal by graphite oxide. *Journal of Water Process Engineering*, 2022. **49**: p. 103085.
44. Perreault, F., A.F. De Faria, and M. Elimelech, Environmental applications of graphene-based nanomaterials. *Chemical Society Reviews*, 2015. **44**(16): p. 5861–5896.

45. Anandhi, P., et al., The enhanced energy density of rGO/TiO2 based nanocomposite as electrode material for supercapacitor. 2022. **11**(11): p. 1792.
46. Yi, H., et al., Environment-friendly fullerene separation methods. *Chemical Engineering Journal*, 2017. **330**: p. 134–145.
47. Yadav, J., Fullerene: properties, synthesis and application. *Research and Reviews: Journal of Physics*, 2017. **6**(3): p. 1–6.
48. Baby, R., B. Saifullah, and M.Z. Hussein, Carbon Nanomaterials for the Treatment of Heavy Metal-Contaminated Water and Environmental Remediation. *Nanoscale Research Letters*, 2019. **14**(1): p. 341.
49. Qiu, X., et al., Applications of nanomaterials in asymmetric photocatalysis: recent progress, challenges, and opportunities. *Advanced Materials*, 2021. **33**(6): p. 2001731.
50. Dang, S., Q.-L. Zhu, and Q. Xu, Nanomaterials derived from metal–organic frameworks. *Nature Reviews Materials*, 2017. **3**(1): p. 17075.
51. Khan, S.A., Chapter 1 – Metal nanoparticles toxicity: Role of physicochemical aspects, in *Metal Nanoparticles for Drug Delivery and Diagnostic Applications*, M.R. Shah, M. Imran, and S. Ullah, Editors. 2020, Elsevier. p. 1–11.
52. Mitra, A. and G. De, Chapter 6 – Sol-Gel synthesis of metal nanoparticle incorporated oxide films on glass, in *Glass Nanocomposites*, B. Karmakar, K. Rademann, and A.L. Stepanov, Editors. 2016, William Andrew Publishing: Boston. p. 145–163.
53. Jamkhande, P.G., et al., Metal nanoparticles synthesis: An overview on methods of preparation, advantages and disadvantages, and applications. *Journal of Drug Delivery Science and Technology*, 2019. **53**: p. 101174.
54. He, L., et al., Synthesis, characterization, and application of metal nanoparticles supported on nitrogen-doped carbon: Catalysis beyond electrochemistry. *Angewandte Chemie International Edition*, 2016. **55**(41): p. 12582–12594.
55. Mahmoud, A.E.D., recent advances of TiO2 nanocomposites for photocatalytic degradation of water contaminants and rechargeable sodium ion batteries, in *Advances in Nanocomposite Materials for Environmental and Energy Harvesting Applications*, A.E. Shalan, A.S. Hamdy Makhlouf, and S. Lanceros-Méndez, Editors. 2022, Springer International Publishing: Cham. p. 757–770.
56. Falcaro, P., et al., Application of metal and metal oxide nanoparticles@MOFs. *Coordination Chemistry Reviews*, 2016. **307**: p. 237–254.
57. Jeevanandam, J., Y.S. Chan, and M.K. Danquah, Biosynthesis of Metal and Metal Oxide Nanoparticles. *ChemBioEng Reviews*, 2016. **3**(2): p. 55–67.
58. Nikam, A., B. Prasad, and A. Kulkarni, Wet chemical synthesis of metal oxide nanoparticles: A review. *CrystEngComm*, 2018. **20**(35): p. 5091–5107.
59. Terna, A.D., et al., The future of semiconductors nanoparticles: Synthesis, properties and applications. *Materials Science and Engineering: B*, 2021. **272**: p. 115363.
60. Feng, Y., et al., Biosynthetic transition metal chalcogenide semiconductor nanoparticles: Progress in synthesis, property control and applications. *Current Opinion in Colloid & Interface Science*, 2018. **38**: p. 190–203.
61. Bajorowicz, B., et al., Quantum dot-decorated semiconductor micro-and nanoparticles: A review of their synthesis, characterization and application in photocatalysis. *Advances in Colloid and Interface Science*, 2018. **256**: p. 352–372.
62. Nasir, A., A. Kausar, and A. Younus, A review on preparation, properties and applications of polymeric nanoparticle-based materials. *Polymer-Plastics Technology and Engineering*, 2015. **54**(4): p. 325–341.
63. Crucho, C.I.C. and M.T. Barros, Polymeric nanoparticles: A study on the preparation variables and characterization methods. *Materials Science & Engineering C-Materials for Biological Applications*, 2017. **80**: p. 771–784.

64. Janaszewska, A., et al., Cytotoxicity of Dendrimers. *Biomolecules*, 2019. **9**(8): p. 330.
65. Mondal, S., Review on nanocellulose polymer nanocomposites. *Polymer-Plastics Technology and Engineering*, 2018. **57**(13): p. 1377–1391.
66. Kim, J.-H., et al., Review of nanocellulose for sustainable future materials. *International Journal of Precision Engineering and Manufacturing-Green Technology*, 2015. **2**(2): p. 197–213.
67. Singh, D., et al., Ceramic nanoparticles: Recompense, cellular uptake and toxicity concerns. *Artificial Cells, Nanomedicine, and Biotechnology*, 2016. **44**(1): p. 401–409.
68. Baaloudj, O., et al., A comparative study of ceramic nanoparticles synthesized for antibiotic removal: Catalysis characterization and photocatalytic performance modeling. *Environmental Science and Pollution Research*, 2021. **28**(11): p. 13900–13912.
69. Thomas, S.C., P. Kumar Mishra, and S. Talegaonkar, Ceramic nanoparticles: fabrication methods and applications in drug delivery. *Current Pharmaceutical Design*, 2015. **21**(42): p. 6165–6188.
70. Balasubramanian, S., B. Gurumurthy, and A. Balasubramanian, Biomedical applications of ceramic nanomaterials: A review. *International Journal of Pharmaceutical Sciences and Research*, 2017. **8**(12): p. 4950–4959.
71. Wu, H., et al., Electrospinning of ceramic nanofibers: Fabrication, assembly and applications. *Journal of Advanced Ceramics*, 2012. **1**(1): p. 2–23.
72. Omanović-Mikličanin, E., et al., Nanocomposites: A brief review. *Health and Technology*, 2020. **10**(1): p. 51–59.
73. Ates, B., et al., Chemistry, structures, and advanced applications of nanocomposites from biorenewable resources. *Chemical Reviews*, 2020. **120**(17): p. 9304–9362.
74. Fu, S., et al., Some basic aspects of polymer nanocomposites: A critical review. *Nano Materials Science*, 2019. **1**(1): p. 2–30.
75. Bhat, A., et al., Review on nanocomposites based on aerospace applications. *Nanotechnology Reviews*, 2021. **10**(1): p. 237–253.
76. Sonawane, G.H., S.P. Patil, and S.H. Sonawane, Nanocomposites and its applications, in *Applications of Nanomaterials*. 2018, Elsevier. p. 1–22.
77. Kusdianto, K., et al., Fabrication of TiO2–Ag nanocomposite thin films via one-step gas-phase deposition. *Ceramics International*, 2017. **43**(6): p. 5351–5355.
78. Zhang, Q., et al., Recent developments in single-walled carbon nanotube thin films fabricated by dry floating catalyst chemical vapor deposition. *Topics in Current Chemistry*, 2017. **375**(6): p. 90.
79. Wirunchit, S., P. Gansa, and W. Koetniyom, Synthesis of ZnO nanoparticles by ball-milling process for biological applications. *Materials Today: Proceedings*, 2021. **47**: p. 3554–3559.
80. Mirzaei, A., et al., Characterization and optical studies of PVP-capped silver nanoparticles. *Journal of Nanostructure in Chemistry*, 2017. **7**(1): p. 37–46.
81. Fakhari, S., M. Jamzad, and H. Kabiri Fard, Green synthesis of zinc oxide nanoparticles: A comparison. *Green Chemistry Letters and Reviews*, 2019. **12**(1): p. 19–24.
82. Scuderi, M., et al., Nanoscale study of the tarnishing process in electron beam lithography-fabricated silver nanoparticles for plasmonic applications. *The Journal of Physical Chemistry C*, 2016. **120**(42): p. 24314–24323.
83. Krishnia, L., P. Thakur, and A. Thakur, Synthesis of nanoparticles by physical route, in *Synthesis and Applications of Nanoparticles*, A. Thakur, P. Thakur, and S.M.P. Khurana, Editors. 2022, Springer Nature Singapore: Singapore. p. 45–59.
84. Iravani, S., et al., Synthesis of silver nanoparticles: Chemical, physical and biological methods. *Research in Pharmaceutical Sciences*, 2014. **9**(6): p. 385–406.
85. Grammatikopoulos, P., et al., Nanoparticle design by gas-phase synthesis. *Advances in Physics: X*, 2016. **1**(1): p. 81–100.

86. Benetti, G., et al., Antimicrobial nanostructured coatings: A gas-phase deposition and magnetron sputtering perspective. *Materials*, 2020. **13**(3): p. 784.

87. Guo, J., et al., Suppressing the photocatalytic activity of TiO2 nanoparticles by extremely thin Al$_2$O$_3$ films grown by gas-phase deposition at ambient conditions. *Nanomaterials*, 2018. **8**(2): p. 61.

88. Pala, N. and M. Karabiyik, Electron Beam Lithography (EBL), in *Encyclopedia of Nanotechnology*, B. Bhushan, Editor. 2016, Springer Netherlands: Dordrecht. p. 1033–1057.

89. Wang, Y., et al., Direct wavelength-selective optical and electron-beam lithography of functional inorganic nanomaterials. *ACS Nano*, 2019. **13**(12): p. 13917–13931.

90. Wu, T. and Y.-W. Lin, Surface-enhanced Raman scattering active gold nanoparticle/nanohole arrays fabricated through electron beam lithography. *Applied Surface Science*, 2018. **435**: p. 1143–1149.

91. Mahmoud, A.E.D., et al., Mechanochemical versus chemical routes for graphitic precursors and their performance in micropollutants removal in water. *Powder Technology*, 2020. **366**: p. 629–640.

92. Kumar, M., et al., Ball milling as a mechanochemical technology for fabrication of novel biochar nanomaterials. *Bioresource Technology*, 2020. **312**: p. 123613.

93. El-Eskandarany, M.S., et al., Mechanical milling: A superior nanotechnological tool for fabrication of nanocrystalline and nanocomposite materials. *Nanomaterials*, 2021. **11**(10): p. 2484.

94. Balamurugan, S., et al. ZnO nanoparticles obtained by ball milling technique: Structural, micro-structure, optical and photo-catalytic properties. in AIP Conference Proceedings. 2016. AIP Publishing LLC.

95. Yogesh, G.K., et al., Progress in pulsed laser ablation in liquid (PLAL) technique for the synthesis of carbon nanomaterials: A review. *Applied Physics A*, 2021. **127**(11): p. 1–40.

96. Fazio, E., et al., Nanoparticles engineering by pulsed laser ablation in liquids: Concepts and applications. *Nanomaterials*, 2020. **10**(11): p. 2317.

97. Mostafa, A.M., et al., Synthesis of cadmium oxide nanoparticles by pulsed laser ablation in liquid environment. *Optik*, 2017. **144**: p. 679–684.

98. Prabhu, S. and E.K. Poulose, Silver nanoparticles: Mechanism of antimicrobial action, synthesis, medical applications, and toxicity effects. *International Nano Letters*, 2012. **2**(1): p. 32.

99. Heuer-Jungemann, A., et al., The role of ligands in the chemical synthesis and applications of inorganic nanoparticles. *Chemical Reviews*, 2019. **119**(8): p. 4819–4880.

100. Khan, M., et al., Plant extracts as green reductants for the synthesis of silver nanoparticles: Lessons from chemical synthesis. *Dalton Transactions*, 2018. **47**(35): p. 11988–12010.

101. Ram, J., et al., Effect of Annealing on the Surface Morphology, Optical and and Structural Properties of Nanodimensional Tungsten Oxide Prepared by Coprecipitation Technique. *Journal of Electronic Materials*, 2019. **48**(2): p. 1174–1183.

102. Razanajatovo, M.R., et al., Selective adsorption of phosphate in water using lanthanum-based nanomaterials: A critical review. *Chinese Chemical Letters*, 2021. **32**(9): p. 2637–2647.

103. Pachiyappan, J., N. Gnanasundaram, and G.L. Rao, Preparation and characterization of ZnO, MgO and ZnO–MgO hybrid nanomaterials using green chemistry approach. *Results in Materials*, 2020. **7**: p. 100104.

104. James, M., et al., Microfluidic synthesis of iron oxide nanoparticles. *Nanomaterials*, 2020. **10**(11): p. 2113.

105. Modan, E.M. and A.G. Plăiașu, Advantages and disadvantages of chemical methods in the elaboration of nanomaterials. *The Annals of "Dunarea de Jos" University of Galati. Fascicle IX, Metallurgy and Materials Science*, 2020. **43**(1): p. 53–60.

106. Richard, B., J.-L. Lemyre, and A.M. Ritcey, Nanoparticle size control in microemulsion synthesis. *Langmuir*, 2017. **33**(19): p. 4748–4757.
107. Daruich De Souza, C., B. Ribeiro Nogueira, and M.E.C.M. Rostelato, Review of the methodologies used in the synthesis gold nanoparticles by chemical reduction. *Journal of Alloys and Compounds*, 2019. **798**: p. 714–740.
108. López-Naranjo, E.J., et al., Transparent Electrodes: A Review of the Use of Carbon-Based Nanomaterials. *Journal of Nanomaterials*, 2016. **2016**: p. 4928365.
109. Shandilya, M., R. Rai, and J. Singh, Hydrothermal technology for smart materials. *Advances in Applied Ceramics*, 2016. **115**(6): p. 354–376.
110. Darr, J.A., et al., Continuous hydrothermal synthesis of inorganic nanoparticles: Applications and future directions. *Chemical Reviews*, 2017. **117**(17): p. 11125–11238.
111. Gan, Y.X., et al., Hydrothermal Synthesis of Nanomaterials. *Journal of Nanomaterials*, 2020. **2020**: p. 8917013.
112. Parveen, K., V. Banse, and L. Ledwani. Green synthesis of nanoparticles: Their advantages and disadvantages. in AIP conference proceedings. 2016. AIP Publishing LLC.
113. Mirzaei, H. and M. Darroudi, Zinc oxide nanoparticles: Biological synthesis and biomedical applications. *Ceramics International*, 2017. **43**(1): p. 907–914.
114. Mahmoud, A.E.D., Eco-friendly reduction of graphene oxide via agricultural byproducts or aquatic macrophytes. *Materials Chemistry and Physics*, 2020. **253**: p. 123336.
115. Omran, M.A., et al., Optimization of mild steel corrosion inhibition by water hyacinth and common reed extracts in acid media using factorial experimental design. *Green Chemistry Letters and Reviews*, 2022. **15**(1): p. 216–232.
116. Lee, S.H. and B.-H. Jun, Silver nanoparticles: Synthesis and application for nanomedicine. *International Journal of Molecular Sciences*, 2019. **20**(4): p. 865.
117. Bachheti, R.K., et al., Chapter 22 – Algae-, fungi-, and yeast-mediated biological synthesis of nanoparticles and their various biomedical applications, in *Handbook of Greener Synthesis of Nanomaterials and Compounds*, B. Kharisov and O. Kharissova, Editors. 2021, Elsevier. p. 701–734.
118. Mahmoud, A.E.D., et al., Green copper oxide nanoparticles for lead, nickel, and cadmium removal from contaminated water. *Scientific Reports*, 2021. **11**(1): p. 12547.
119. Hasan, S., A review on nanoparticles: Their synthesis and types. *Research Journal of Recent Sciences*, 2015. **2277**: p. 2502.
120. Balakumaran, M., R. Ramachandran, and P. Kalaichelvan, Exploitation of endophytic fungus, Guignardia mangiferae for extracellular synthesis of silver nanoparticles and their in vitro biological activities. *Microbiological Research*, 2015. **178**: p. 9–17.
121. Abdeen, M., et al., Microbial-Physical Synthesis of Fe and Fe_3O_4 Magnetic Nanoparticles Using *Aspergillus niger* YESM1 and Supercritical Condition of Ethanol. *Journal of Nanomaterials*, 2016. **2016**: p. 9174891.
122. Nikolaidis, P., Analysis of Green Methods to Synthesize Nanomaterials, in *Green Synthesis of Nanomaterials for Bioenergy Applications*. 2020. p. 125–144.
123. Singh, P., et al., Biological Synthesis of Nanoparticles from Plants and Microorganisms. *Trends in Biotechnology*, 2016. **34**(7): p. 588–599.
124. Hosny, M., et al., Comparative study between Phragmites australis root and rhizome extracts for mediating gold nanoparticles synthesis and their medical and environmental applications. *Advanced Powder Technology*, 2021. **32**(7): p. 2268–2279.
125. Wang, R., et al., Critical problems faced in Raman-based energy transport characterization of nanomaterials. *Physical Chemistry Chemical Physics*, 2022. **24**. p: 22390–22404.
126. Malekpour, H. and A.A. Balandin, Raman-based technique for measuring thermal conductivity of graphene and related materials. *Journal of Raman Spectroscopy*, 2018. **49**(1): p. 106–120.

127. Xu, X., et al., Near-field enhanced plasmonic-magnetic bifunctional nanotubes for single cell bioanalysis. *Advanced Functional Materials*, 2013. **23**(35): p. 4332–4338.

128. Bruzas, I., et al., Advances in surface-enhanced Raman spectroscopy (SERS) substrates for lipid and protein characterization: Sensing and beyond. *Analyst*, 2018. **143**(17): p. 3990–4008.

129. Mudalige, T., et al., Chapter 11 – Characterization of Nanomaterials: Tools and Challenges, in *Nanomaterials for Food Applications*, A. López Rubio, et al., Editors. 2019, Elsevier. p. 313–353.

130. Tang, B., H. Guoxin, and H. Gao, Raman Spectroscopic Characterization of Graphene. *Applied Spectroscopy Reviews*, 2010. **45**(5): p. 369–407.

131. Zhao, B., et al., A green, facile, and rapid method for microextraction and Raman detection of titanium dioxide nanoparticles from milk powder. *RSC Advances*, 2017. **7**(35): p. 21380–21388.

132. Gupta, J., P. Bhargava, and D. Bahadur, Morphology dependent photocatalytic and magnetic properties of ZnO nanostructures. *Physica B: Condensed Matter*, 2014. **448**: p. 16–19.

133. Salame, P.H., V.B. Pawade, and B.A. Bhanvase, Chapter 3 – Characterization Tools and Techniques for Nanomaterials, in *Nanomaterials for Green Energy*, B.A. Bhanvase, et al., Editors. 2018, Elsevier. p. 83–111.

134. Lu, P.-J., et al., Analysis of titanium dioxide and zinc oxide nanoparticles in cosmetics. *Journal of Food and Drug Analysis*, 2015. **23**(3): p. 587–594.

135. Vladár, A.E. and V.-D. Hodoroaba, Chapter 2.1.1 – Characterization of nanoparticles by scanning electron microscopy, in *Characterization of Nanoparticles*, V.-D. Hodoroaba, W.E.S. Unger, and A.G. Shard, Editors. 2020, Elsevier. p. 7–27.

136. Badr, N.B.E., K.M. Al-Qahtani, and A.E.D. Mahmoud, Factorial experimental design for optimizing selenium sorption on Cyperus laevigatus biomass and green-synthesized nano-silver. *Alexandria Engineering Journal*, 2020. **59**(6): p. 5219–5229.

137. Akhtar, K., et al., Scanning electron microscopy: Principle and applications in nanomaterials characterization, in *Handbook of Materials Characterization*. 2018, Springer. p. 113–145.

138. Motas, J.G., et al., XPS, SEM, DSC and Nanoindentation Characterization of Silver Nanoparticle-Coated Biopolymer Pellets. *Applied Sciences*, 2021. **11**(16): p. 7706.

139. Lu, P.-J., et al., Methodology for sample preparation and size measurement of commercial ZnO nanoparticles. *Journal of Food and Drug Analysis*, 2018. **26**(2): p. 628–636.

140. Pentassuglia, S., V. Agostino, and T. Tommasi, EAB—Electroactive Biofilm: A Biotechnological Resource, in *Encyclopedia of Interfacial Chemistry*, K. Wandelt, Editor. 2018, Elsevier: Oxford. p. 110–123.

141. NicDaéid, N., Forensic Sciences | Systematic Drug Identification☆, in *Encyclopedia of Analytical Science (Third Edition)*, P. Worsfold, et al., Editors. 2019, Academic Press: Oxford. p. 75–80.

142. Mourdikoudis, S., R.M. Pallares, and N.T.K. Thanh, Characterization techniques for nanoparticles: Comparison and complementarity upon studying nanoparticle properties. *Nanoscale*, 2018. **10**(27): p. 12871–12934.

143. Madakka, M., N. Jayaraju, and N. Rajesh, Mycosynthesis of silver nanoparticles and their characterization. *MethodsX*, 2018. **5**: p. 20–29.

144. Venkatachalam, S., Chapter 6 – Ultraviolet and visible spectroscopy studies of nanofillers and their polymer nanocomposites, in *Spectroscopy of Polymer Nanocomposites*, S. Thomas, D. Rouxel, and D. Ponnamma, Editors. 2016, William Andrew Publishing. p. 130–157.

145. Begum, R., et al., Applications of UV/Vis spectroscopy in characterization and catalytic activity of noble metal nanoparticles fabricated in responsive polymer microgels: A review. *Critical Reviews in Analytical Chemistry*, 2018. **48**(6): p. 503–516.

146. Zhang, H., J. Jung, and Y. Zhao, Preparation, characterization and evaluation of antibacterial activity of catechins and catechins–Zn complex loaded β-chitosan nanoparticles of different particle sizes. *Carbohydrate Polymers*, 2016. **137**: p. 82–91.

147. Talari, A.C.S., et al., Advances in Fourier transform infrared (FTIR) spectroscopy of biological tissues. *Applied Spectroscopy Reviews*, 2017. **52**(5): p. 456–506.

148. Faghihzadeh, F., et al., Fourier transform infrared spectroscopy to assess molecular-level changes in microorganisms exposed to nanoparticles. *Nanotechnology for Environmental Engineering*, 2016. **1**(1): p. 1–16.

149. Dincer, I. and C. Acar, Review and evaluation of hydrogen production methods for better sustainability. *International Journal of Hydrogen Energy*, 2015. **40**(34): p. 11094–11111.

150. Kumar, G., et al., Application of nanotechnology in dark fermentation for enhanced biohydrogen production using inorganic nanoparticles. *International Journal of Hydrogen Energy*, 2019. **44**(26): p. 13106–13113.

151. Zaidi, A.A., et al., Nanoparticles augmentation on biogas yield from microalgal biomass anaerobic digestion. *International Journal of Hydrogen Energy*, 2018. **43**(31): p. 14202–14213.

152. Hussain, C.M., *Handbook of Functionalized Nanomaterials for Industrial Applications*. 2020: Elsevier.

153. Gadhe, A., S.S. Sonawane, and M.N. Varma, Enhancement effect of hematite and nickel nanoparticles on biohydrogen production from dairy wastewater. *International Journal of Hydrogen Energy*, 2015. **40**(13): p. 4502–4511.

154. Ganguly, P., et al., 2D Nanomaterials for Photocatalytic Hydrogen Production. *ACS Energy Letters*, 2019. **4**(7): p. 1687–1709.

155. Hwang, H.T. and A. Varma, Hydrogen storage for fuel cell vehicles. *Current Opinion in Chemical Engineering*, 2014. **5**: p. 42–48.

156. Boateng, E. and A. Chen, Recent advances in nanomaterial-based solid-state hydrogen storage. *Materials Today Advances*, 2020. **6**: p. 100022.

157. Pasquini, L., Design of nanomaterials for hydrogen storage. *Energies*, 2020. **13**(13): p. 3503.

158. Niaz, S., T. Manzoor, and A.H. Pandith, Hydrogen storage: Materials, methods and perspectives. *Renewable and Sustainable Energy Reviews*, 2015. **50**: p. 457–469.

159. Hassan, M., et al., 2 – Bimetallic and trimetallic nanomaterials for hydrogen storage applications, in *Nanomaterials for Hydrogen Storage Applications*, F. Sen, A. Khan, and A.M. Asiri, Editors. 2021, Elsevier. p. 17–36.

160. Mahmoud, A.E.D., et al., Environmental bioremediation as an eco-sustainable approach for pesticides: A case study of MENA region, in *Pesticides Bioremediation*. 2022, Springer. p. 479.

161. Ziarati, P., et al., Bioremediation of pharmaceutical effluent by food industry and agricultural waste biomass, in *Hospital Wastewater Treatment: Global Scenario and Case Studies*, N.A. Khan, et al., Editors. 2022, IWA Publishing.

162. Mahmoud, A.E.D., et al., 26 – Water resources security and management for sustainable communities, in *Phytochemistry, the Military and Health*, A.G. Mtewa and C. Egbuna, Editors. 2021, Elsevier. p. 509–522.

163. El-Sayed, M.E.A., Nanoadsorbents for water and wastewater remediation. *Science of the Total Environment*, 2020. **739**: p. 139903.

164. Vambol, S., et al., Comprehensive insights into sources of pharmaceutical wastewater in the biotic systems, in *Pharmaceutical Wastewater Treatment Technologies: Concepts and Implementation Strategies*, N.A. Khan, et al., Editors. 2021, IWA Publishing. p. 0.

165. Mahmoud, A.E.D. and S. Kathi, Chapter 7 – Assessment of biochar application in decontamination of water and wastewater, in *Cost Effective Technologies for Solid Waste and Wastewater Treatment*, S. Kathi, S. Devipriya, and K. Thamaraiselvi, Editors. 2022, Elsevier. p. 69–74.
166. El Din Mahmoud, A. and M. Fawzy, Bio-based methods for wastewater treatment: Green sorbents, in *Phytoremediation: Management of Environmental Contaminants*, Volume 3, A.A. Ansari, et al., Editors. 2016, Springer International Publishing: Cham. p. 209–238.
167. Lu, H., et al., An overview of nanomaterials for water and wastewater treatment. *Advances in Materials Science and Engineering*, 2016. **2016**: p. 4964828.
168. Amin, M.T., A.A. Alazba, and U. Manzoor, A Review of Removal of Pollutants from Water/Wastewater Using Different Types of Nanomaterials. *Advances in Materials Science and Engineering*, 2014. **2014**: p. 825910.
169. Yin, Z., et al., The application of carbon nanotube/graphene-based nanomaterials in wastewater treatment. *Small*, 2020. **16**(15): p. 1902301.
170. Zhang, C., X. Chen, and S.-H. Ho, Wastewater treatment nexus: Carbon nanomaterials towards potential aquatic ecotoxicity. *Journal of Hazardous Materials*, 2021. **417**: p. 125959.
171. Xue, Z., et al., Adsorption of Cd (II) in water by mesoporous ceramic functional nanomaterials. *Royal Society Open Science*, 2019. **6**(4): p. 182195.
172. Vauthier, C. and G. Ponchel, *Polymer Nanoparticles for Nanomedicines*. 2017: Springer.
173. Kajani, A.A., et al., Anticancer effects of silver nanoparticles encapsulated by Taxus baccata extracts. *Journal of Molecular Liquids*, 2016. **223**: p. 549–556.
174. Sadat Shandiz, S.A., et al., Novel imatinib-loaded silver nanoparticles for enhanced apoptosis of human breast cancer MCF-7 cells. *Artificial Cells, Nanomedicine, and Biotechnology*, 2017. **45**(6): p. 1082–1091.
175. Kong, F.-Y., et al., Unique roles of gold nanoparticles in drug delivery, targeting and imaging applications. *Molecules*, 2017. **22**(9): p. 1445.
176. Wang, W., et al., Recent applications of nanomaterials in prosthodontics. *Journal of Nanomaterials*, 2015. **2015**.
177. Hamdan, S., et al., Nanotechnology-driven therapeutic interventions in wound healing: Potential uses and applications. *ACS Central Science*, 2017. **3**(3): p. 163–175.
178. Borrego, B., et al., Potential application of silver nanoparticles to control the infectivity of Rift Valley fever virus in vitro and in vivo. *Nanomedicine: Nanotechnology, Biology and Medicine*, 2016. **12**(5): p. 1185–1192.
179. Yezdani, U., et al., Application of nanotechnology in diagnosis and treatment of various diseases and its future advances in medicine. *World Journal of Pharmacy and Pharmaceutical Sciences*, 2018. **7**(11): p. 1611–1633.
180. Maiti, D., et al., Carbon-based nanomaterials for biomedical applications: A recent study. *Frontiers in Pharmacology*, 2019. **9**: p. 1401.
181. Sharma, N., et al., Preparation and catalytic applications of nanomaterials: A review. *RSC Advances*, 2015. **5**(66): p. 53381–53403.
182. Liu, Z., et al., Sulfonated magnetic carbon nanotube arrays as effective solid acid catalysts for the hydrolyses of polysaccharides in crop stalks. *Catalysis Communications*, 2014. **56**: p. 1–4.
183. Zhu, Q.-L. and Q. Xu, Immobilization of ultrafine metal nanoparticles to high-surface-area materials and their catalytic applications. *Chem*, 2016. **1**(2): p. 220–245.
184. Nasrollahzadeh, M., et al., Immobilization of copper nanoparticles on perlite: Green synthesis, characterization and catalytic activity on aqueous reduction of 4-nitrophenol. *Journal of Molecular Catalysis A: Chemical*, 2015. **400**: p. 22–30.

185. Yang, F., et al., Understanding nano effects in catalysis. *National Science Review*, 2015. **2**(2): p. 183–201.

186. Shapira, P. and J. Youtie, The economic contributions of nanotechnology to green and sustainable growth, in *Green Processes for Nanotechnology*. 2015, Springer. p. 409–434.

187. Soni, R.A., M. Rizwan, and S. Singh, Opportunities and potential of green chemistry in nanotechnology. *Nanotechnology for Environmental Engineering*, 2022. **7**: p. 661–673.

188. Polshettiwar, V. and R.S. Varma, Green chemistry by nano-catalysis. *Green Chemistry*, 2010. **12**(5): p. 743–754.

189. Carpenter, M.A., S. Mathur, and A. Kolmakov, *Metal Oxide Nanomaterials for Chemical Sensors*. 2012: Springer Science & Business Media.

190. Chaudhary, S., et al., Chemical sensing applications of ZnO nanomaterials. *Materials*, 2018. **11**(2): p. 287.

191. Davarnejad, R., et al., A green technique for synthesising iron oxide nanoparticles by extract of centaurea cyanus plant: An optimised adsorption process for methylene blue. *International Journal of Environmental Analytical Chemistry*, 2022. **102**(10): p. 2379–2393.

3 Advances in Thermochemical Hydrogen Production Using Nanomaterials

An Analysis of Production Methods, Challenges, and Opportunities

Emmanuel I. Epelle, Winifred Obande, Ovis D. Irefu, and Alivia Mukherjee

3.1 INTRODUCTION

Hydrogen (H_2) has a tremendous potential to significantly contribute to the world's future energy requirements, and this potential has been recognized by several researchers. A recent study by the Hydrogen Council [1] envisages that up to 18% of the world's energy demands (78 EJ) can be supplied by H_2 in 2050. Its production via clean technologies has been intensely researched over the last two decades, and this is primarily because only <1% of globally produced H_2 is currently from renewable sources [2]. Thus, the development and scale-up of novel and sustainable H_2 production technologies can foster the H_2 economy and the future of sustainable energy. Thermochemical processes for H_2 production typically involve the use of thermal energy to drive a chemical reaction that generates H_2 from a feedstock such as fossil fuels, biomass, or water (Figure 3.1). These processes majorly include natural gas reforming, biomass gasification, biomass-derived liquid reforming, and solar thermochemical H_2 production.

The production of H_2 from fossil fuels (via steam methane reforming – SMR) is the most intensely explored production route, owing to its low production cost; however, given the reliance of this production route on fossil fuels, it is an environmentally harmful process. Conversely, H_2 production from lignocellulosic biomass sources often occurs thermochemically via four main methods – pyrolysis, liquefaction, gasification, and combustion – with minimal net emission of greenhouse gases. The coproduction of methane, carbon monoxide, and other gaseous products during

DOI: 10.1201/9781003371007-3

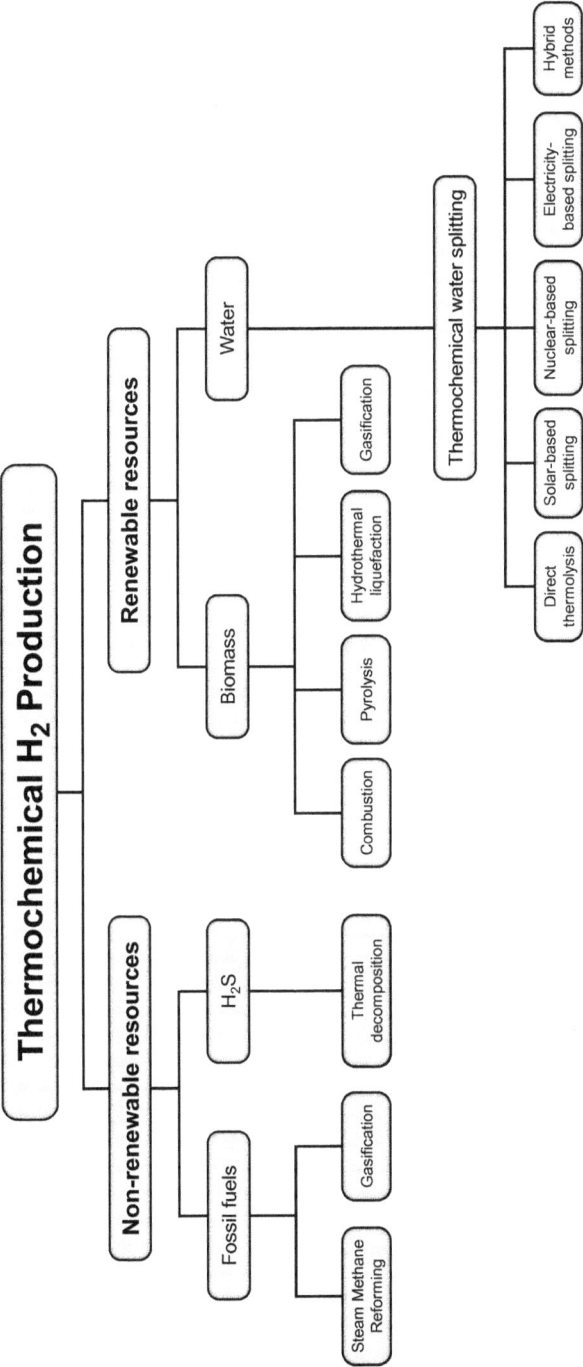

FIGURE 3.1 Classification of thermochemical processes for H_2 production.

gasification and pyrolysis, can be subsequently transformed into H_2 using natural gas reforming and water-gas shift (WGS) reforming methods.

Water splitting, another viable H_2 production route involves the use of energy to break down water into its components (H_2 and O_2); and this approach also offers the benefit of producing little to no greenhouse gas emissions. The driving energy sources for the water-splitting process could be thermal, mechanical, electrical, photonic, or biological. The use of solar energy has been relied on in several water-splitting endeavors; this is a result of its minimal environmental impact, availability, intensity and reliability – it has been estimated that the solar energy captured by the earth for a duration of one hour is sufficient to meet the global energy demand for a year [3]. There have been various techniques developed by researchers to produce H_2 from water using solar energy; these include thermolysis, thermochemical, photodissociation, photoelectrolysis, photoelectrochemical, photochemical, photodecomposition, photocatalytic, photolysis, photodegradation, photobiological, water electrolysis, and hybrid methods [4]. Some other authors have broadly classified them into thermochemical, photobiological, and photocatalytic water splitting [5, 6]. Compared to fuel reforming processes, thermochemical water splitting (TCWS) is a more sustainable route for H_2 generation because it uses abundant and renewable resources like water and solar energy, thereby eliminating the need for fossil fuels, and reducing greenhouse gas emissions.

Nonetheless, the efficiency of the solar-based technologies for TCWS requires improvement. Key to this, is the enhancement of solar collector designs and the development of novel nanocatalysts, required for increased and stable H_2 generation efficiency. Catalytic nanomaterials (nanocatalysts) also play a crucial role in fossil fuel- and biomass-derived H_2. In this chapter, an overview of the applications of nanomaterials for addressing the inefficiencies of different thermochemical H_2 production processes is presented. The mechanisms, challenges and prospects of effective nanomaterial utilization are also outlined to facilitate the design and implementation of lab-scale, pilot-scale, and full industrial-scale thermochemical processes. A more detailed analysis of the thermochemical H_2 production routes is subsequently presented (in Sections 3.2 and 3.3), with special emphasis placed on the thermochemical water-splitting cycles (Sections 3.4 and 3.5), given their immense potential.

3.2   NATURAL GAS REFORMING

A huge portion (95%) of the commercially produced H_2 in the US today, is from steam methane reforming (SMR) [7] (Equation 3.1). In addition, SMR is a mature technology involving the catalytic application of high-temperature steam (700°C–1000°C) to produce H_2 from a methane source with pressure conditions in the range of 3–25 bar [8]. Carbon monoxide (CO) is also produced in considerable quantities together with a low amount of carbon dioxide (CO_2). The CO is subsequently reacted with steam via a catalyst to produce CO_2 and more H_2 (the water-gas shift reaction – Equation 3.2). A series of separation stages (usually adsorption-based processes) is conducted, leaving pure H_2 as the product.

There have been several developments in the application of nanomaterial-based catalysts for the improvement of SMR operations [9]. These improvements have taken the form of overall temperature reduction, increased methane to H_2 conversion, induction heating, reduced coke formation, and increased cost effectiveness. It is worth mentioning that catalyst deactivation, which mainly occurs via coke deposition on the surface of conventional catalysts, is a severe drawback of SMR [10]; this negatively impacts the long-term stability of the utilized catalyst and consequently the operational costs – a challenge which the application of nanomaterials can also mitigate.

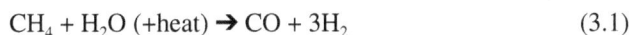

$$CH_4 + H_2O \text{ (+heat)} \rightarrow CO + 3H_2 \tag{3.1}$$

$$CO + H_2O \rightarrow CO_2 + H_2 \tag{3.2}$$

$$CH_4 + \tfrac{1}{2}H_2O \rightarrow CO + 2H_2 \text{ (+heat)} \tag{3.3}$$

Lai et al. [11] proposed hybrid nanostructures (Ni – Al_2O_3, Ni – CeO_2 – Al_2O_3) as nanocatalysts for SMR. They developed a facile synthesis route which uses the Al_2O_3 nanoparticle cluster as a support material and CeO_2 nanoparticles as a promoter to the Ni-based catalyst. Their hybridized catalyst resulted in a low starting temperature (400°C), excellent catalytic activity at this relatively low temperature, increased stability (8 h reaction), a high H_2 yield (3× the equilibrium methane conversion), and reduced coke formation. Another recent development by Vinum et al. [12] involved the tailored synthesis of binary and ternary alloy nanoparticles (CoNi and CuCCoNi) for use as magnetic susceptors for induction heating (the sole energy source) and as catalysts for the steam reforming of methane. They demonstrated the inertness of the developed catalyst to degradation in the first 300 h of operation at reforming conditions (800°C). The study highlights the potential for induction heating in SMR to eliminate costly waste-heat sections applied in traditional SMR setups. This can reduce the design complexity of SMR processes and lower CO_2 emissions. Ligthart et al. [13] demonstrated the crucial influence of particle size (1–9 nm) on the activity and stability of supported Rh nanocatalysts applicable to SMR processes. It was realized that nanoparticles smaller than 2.5 nm are more readily deactivated than larger-sized nanoparticles. The observed deactivation was attributed to the oxidation of these particles under SMR reaction conditions.

Methane and other hydrocarbons present in natural gas may also be partially oxidized (using a limited amount of O_2) for H_2 production (Equation 3.3). This reaction produces CO, which subsequently undergoes the water-shift reaction shown in Equation 3.2. Compared to SMR, which is endothermic, partial oxidation (PO) is an exothermic process, which typically proceeds at a faster rate. However, less H_2 is produced in PO per unit volume of reactant fuel. In the work of De Maron et al. [14] a novel process involving the substitution of SMR with oxy-reforming (i.e. the coupling of SMR with catalytic PO) was proposed. This process enabled better heat exchange, given the endothermic and exothermic attributes of SMR and PO, respectively. A tailored nanocatalyst (Rh-$Ce_{0.5}Zr_{0.5}O_2$) was developed for this purpose. It was reported that the application of this high-activity nanomaterial enhanced methane conversion by promoting methane dissociation on the Ce-based support.

It is evident from Equations 3.1 to 3.3 that natural gas reforming processes release CO_2; additionally, the dependence on fossil fuels implies that it is not a green/

sustainable H_2 production route. Nonetheless, the characteristic low H_2 production cost ($1.8/kg H_2) and its high efficiency (~74%) [15] have made it the subject of several research endeavors, with novel methods proposed to mitigate the associated CO_2 footprint of this process, as earlier highlighted. It is worth mentioning that the thermal decomposition of H_2S is another non-renewable thermochemical H_2 production route (Figure 3.1), for which the reliance on nanomaterials is somewhat limited. The interested reader is referred to the work by Ghahraloud et al. [16], which provides a detailed assessment of this process via a heterogeneous alumina catalyst. With these ongoing advancements, several cleaner production routes are being commensurately developed for H_2 production. Biomass sources, as covered next, represent a crucial H_2 production route for which the application of nanomaterials has been significantly relied upon.

3.3   BIOMASS GASIFICATION AND BIOMASS-DERIVED LIQUID REFORMING

Approximately 80% of the renewable energy supply comes from biomass (such as wood, wood-based residues, landfills, used oils, nonedible oils and municipal waste) [17]. The ability to store biomass and utilize it as needed makes it an attractive source for thermochemical H_2 production. Photosynthesis is the main biomass-producing process, with up to ~720 billion tons of biomass obtained annually through this route [18]. Since the growth of biomass removes CO_2 from the atmosphere, the net carbon emission of this method is usually low, despite biogenic CO_2 production associated with several thermochemical biomass-to-H_2 conversion processes. The net carbon footprint of these conversion processes can be further lowered when they are coupled with long-term carbon capture, utilization, and storage technologies.

Thermochemical routes for biomass conversion to H_2 are majorly carried out via gasification (which occurs with a limited/controlled supply of O_2) and pyrolysis (where no O_2 is included). These differ from combustion in which O_2 supply is typically greater than the stoichiometric requirement, thereby producing CO_2 and water. Compared to conventional heterogeneous catalysts utilized for thermochemical H_2 production from biomass, nanocatalysts mitigate challenges such as mass transfer resistance, inefficient biomass conversion, high-temperature requirements, and rapid deactivation. According to Karimi-Maleh et al. [18] novel advancements in nanocatalysis have further led to the intensification of biomass-based H_2 production process, improvements in gas product quality, improved biomass conversion efficiency, and a significant decrease in tar and char production. A simplified reaction of cellulose-based biomass gasification is shown in Equation 3.4. In this simplification, glucose is presented as a substitute for cellulose. Biomass itself is a complex mixture with a highly variable composition, with cellulose being a major component. A more detailed overview of the governing reactions is presented in [19].

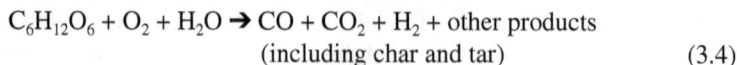

$$C_6H_{12}O_6 + O_2 + H_2O \rightarrow CO + CO_2 + H_2 + \text{other products}$$
$$\text{(including char and tar)} \tag{3.4}$$

Metallic nanomaterials (e.g., Fe, Cu, Pt, Ru, Ni, and Mg) can be used directly as catalysts, supported on other nanomaterials (hybrid nanocatalysts) or impregnated directly to the biomass matrix. Ni nanocatalysts are one of the most intensely explored nanomaterials for thermochemical H_2 production due to their excellent H_2 selectivity under hydrothermal conditions. Nanda et al. [20] impregnated Ni nano-catalysts into lignocellulosic biomasses (wheat straw and pine wood) for enhanced H_2 production via supercritical and subcritical water gasification. The embedding of this catalyst in the lignocellulosic matrix provided several active sites for the enhancement of H_2 yields (2.8–5.8 mmol/g) compared to non-catalytic gasification. According to Norouzi et al. [21], the use of Fe-Ni catalysts supported on γ-Al_2O_3 has been found to enhance catalytic activity during the hydrothermal gasification of algae, achieving up to 12.28 mmol H_2/g of algae. Furthermore, the use of Zn nanoparticles as a promoter for Ru/γ-Al_2O_3 improved the H_2 yield by 16.67% during the hydrothermal gasification of bagasse [22]. Similarly, the inclusion of Cu as a promoter to the Ni/CNT nanocatalyst increased the yield of H_2 by a factor of 5.84 during bagasse gasification as recorded in the work of Rashidi et al. [23].

Fe-Zn/Al_2O_3 nanocatalysts have been employed for the steam reforming of volatiles released during the pyrolysis of wood sawdust [24]. Besides improved H_2 production, a marked resistance to coke formation on the catalyst surface was observed. Additionally, the catalyst exhibited ultra-high stability without sintering – a remarkable cost-saving catalytic H_2 production route. In the work of Tursun et al. [25] olivine and NiO/olivine nanocatalysts were utilized for biomass steam gasification to improve H_2 production and reduce tar production (up to 94% reduction was achieved). Similarly, Li et al. demonstrated the effective application of nano-NiO/Al_2O_3 catalysts for tar removal during sawdust gasification [26]. In their research, Collard et al. [27] discovered that impregnating biomass with an iron-based catalyst resulted in a 20% decrease in tar production and a 43% increase in H_2 gas production compared to a control group without a catalytic reaction. Another study conducted by Shen et al. [28] reported a reduction of heavy tar production by approximately 92.3% during the catalytic pyrolysis of biomass when using a rice husk char-supported nickel-iron nanocatalyst. An increased syngas yield of 2.11 L/g at a temperature of 800°C was also observed. It is worth mentioning that nanocatalysts (particularly those of transition metals) can also be further utilized to convert produced char and tar into H_2 and syngas. It has been demonstrated that up to 99% tar conversion efficiency can be achieved using dolomite, nickel-based and alkali-based nanocatalysts [18, 29]. These successful implementations are indicative of the important and versatile role nano-materials play in biomass-to-H_2 conversion.

Compared to transition metal catalysts, which are prone to sintering and eventual deactivation (resulting from the harsh gasification conditions), metal oxides (e.g. MgO, ZrO_2, ZnO, WO_3, NiO, TiO_2) exhibit better stability and can be used as alternative catalysts for H_2 production from biomass [30, 31]. Some of these are particularly useful for preventing methanation reactions while promoting H_2 formation during biomass gasification, and this is crucial to mitigating the carbon footprint of these processes. In addition to H_2 evolution from the direct thermochemical treatment of biomass, H_2 may also be produced from the accompanying liquids formed during pyrolysis and hydrothermal liquefaction of biomass. Bio-oils and crude bio-ethanol can be reformed to generate H_2 in the presence of a nanocatalyst through a process similar to natural gas reforming. Equation 3.5 represents the steam reforming reaction of bioethanol,

TABLE 3.1
Thermochemical Biomass Conversion to H$_2$ Using Various Catalysts

Biomass	Thermochemical Treatment Method	Nanocatalyst	Size	H$_2$ Yield (%)	Ref.
Pine sawdust	Gasification	NiO/Dolomite	29 nm	70.8%	[33]
Sawdust	Steam gasification	Nano-Ni-La-Fe/γ-Al$_2$O$_3$	–	12.1 wt.%	[34]
Wood sawdust	Catalytic steam reforming	Nano Fe-Zn/Al$_2$O$_3$	–	1.9 wt.%	[24]
Wood sawdust	Pyrolysis and steam reforming	NiO-ZnO-Al$_2$O$_3$	10–11 nm	–	[35]
Bagasse	Steam reforming	Ni-Fe/γ-Al$_2$O$_3$	3.2–4.1 nm	35.9%	[19]
Sugarcane bagasse	Hydrothermal liquefaction and gasification	Cu/γ-Al$_2$O$_3$-MgO	8.4 nm	–	[36]
Municipal solid waste	Steam gasification	Ni-Cu/γ-Al$_2$O$_3$	0.3–0.35 mm *(not in the nano range)*	–	[37]
Tar	Gasification	Olivine, NiO/olivine	0.45–0.90 mm *(not in the nano range)*	55%, 94%	[25]
Tar	Gasification	NiO-Fe$_2$O$_3$/SiO$_2$-γAl$_2$O$_3$	2–50 nm (mesopores): 90–140 nm (macropores)	–	[38]

after which the WGS reaction (Equation 3.2) proceeds to form more H$_2$. It has been shown that Rh-based nanocatalysts (Rh/MgAl$_2$O$_4$, Rh/Y$_2$O$_3$-Al$_2$O$_3$ and RhNi/Y$_2$O$_3$-AlO$_3$) are effective at converting bioethanol (56% 97.2% and 98.2%, respectively) to H$_2$. The controlled preparation of these catalysts as described in [32] led to high H$_2$ selectivity. It was realized that the addition of Ni reduced the deactivation potential, which is partly caused by the numerous impurities typically present in crude bioethanol.

$$C_2H_5OH + H_2O \text{ (+heat)} \rightarrow 2CO + 4H_2 \qquad (3.5)$$

Table 3.1 presents a summary of some nanocatalysts (including their sizes), utilized for H$_2$ production from a variety of biomass sources and the corresponding H$_2$ yields. It can be observed that alumina-based nanocatalysts have been applied for a wide range of biomass sources. It should be mentioned that the successful implementations of larger-sized catalysts are also included to give an indication of the potential for potentially higher H$_2$ yields if nano-sized particles of the same catalyst are utilized.

3.4 THERMOCHEMICAL WATER SPLITTING

Thermochemical cycles for water splitting involve a sequence of chemical reactions that decompose water, using intermediate reactions and substances (auxiliary chemicals) that are all recycled during the process [39]. As a result, the overall reaction is a closed cycle which is equivalent to breaking down water into H_2 and O_2. Theoretically, this process only requires thermal energy as an input, and only water is consumed. Thermochemical cycles can be powered either solely by thermal energy (referred to as pure thermochemical cycles) or by a combination of thermal energy and other forms of energy such as electrical or photonic energy (known as hybrid thermochemical cycles). Although it is possible to directly decompose water into H_2 and O_2 in a single step (without the use of chemicals), the unfavorable thermodynamics and extremely high temperatures required (>2000°C) for this process make it impractical. This temperature is also too high for refractory materials and equipment utilized for such reactions. Additionally, the reaction product is a gas mixture of H_2 and O_2, which poses a significant explosion risk; therefore, direct water thermolysis is not yet a practical engineering route for TCWS. To reduce this temperature requirement, auxiliary chemicals can be introduced.

General Atomic (GA) has identified over 200 thermochemical cycles for water splitting, but only a few of them are capable of large-scale H_2 production and have thus, undergone experimental demonstrations. Two-step thermochemical cycles involve the reduction of a metal oxide in an endothermic step, known as activation, where O_2 is released with the assistance of solar energy. This reduced oxide then reacts with water in an exothermic step, known as hydrolysis, to produce H_2 and the original oxide, which can be recycled to the first step of the cycle (Figure 3.2). A significant advantage of the two-step cycle is its simplicity, in that only two reactions are required; however, the temperature required for the O_2 production reaction typically falls within the range of 1500°C–2500°C, which poses a major challenge in terms of selecting equipment materials that possess desirable characteristics – high solar absorptance, low thermal emittance, good corrosion resistance, and excellent thermal stability. Among the several two-step water-splitting cycles (Fe_3O_4/Fe, CdO/Cd, CeO_2/Ce_2O_3, ZnO/Zn, Co_3O_4/CoO), the ZnO/Zn pairs are considered the most thermodynamically favorable [40]. Figure 3.2 provides a pictorial representation of the two-step cycle.

The application of a three-step cycle (in which the high-temperature step in the two-step process is replaced with a new two-step reaction) has also been utilized for H_2 generation with temperatures below 1500°C realized. The sulfur-iodine (S-I) cycle is the most utilized and extensively studied three-step cycle, with H_2 production rates of up to 10 L/h attained [41]. Additionally, the S-I cycle at the Japan Atomic Energy Agency (JAEA) is capable of reaching a considerably high production rate of 0.065 kg H_2/day [42]. Other established three-step cycles include Fe_3O_4/FeO, CeO_2/Ce_2O_3 and Mn_3O_4/MnO. Additionally, four-step TCWS cycles (including Ce-Cl, Fe-Cl, Cu-Cl, Mg-Cl, and V-Cl) have been proposed, as a means of further reducing the temperature requirements (as low as 550°C for the Cu-Cl cycle) [43]. The V-Cl and Cu-Cl cycles have been reported as the most promising options with efficiencies of up to 46% and 43% attained, respectively [44, 45]. The hybrid sulfur cycle (HyS), otherwise termed the Westinghouse cycle, which uses electricity in addition to thermal energy, is the most popular hybrid cycle for TCWS. Conversion efficiencies as high as 40% have been recorded for this cycle [46]. The key advantage of this cycle is the combined utilization of the ultraviolet (UV) and visible portion of solar radiation for the H_2 generation step.

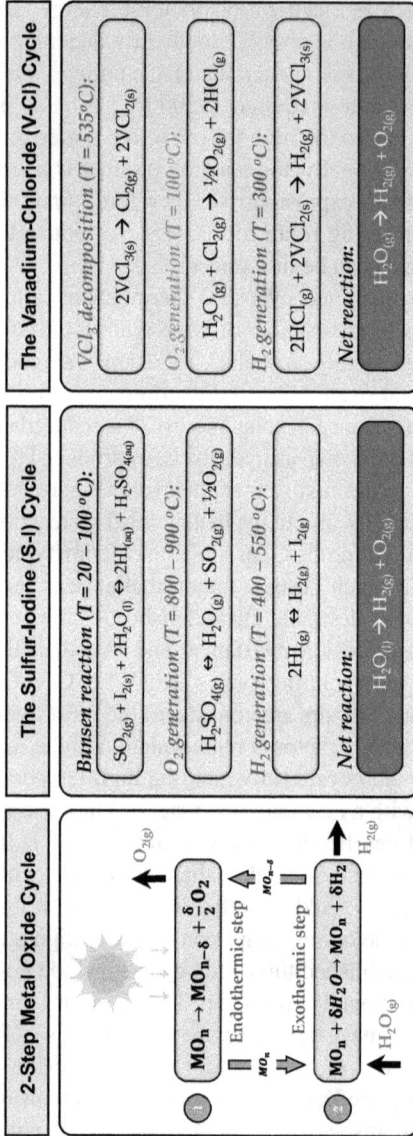

2-Step Metal Oxide Cycle

$$MO_n \rightarrow MO_{n-\delta} + \frac{\delta}{2}O_2$$

Endothermic step

Exothermic step

$$MO_n + \delta H_2O \rightarrow MO_n + \delta H_2$$

$O_{2(g)}$

$MO_{n-\delta}$

MO_n

$H_{2(g)}$

$H_2O_{(g)}$

The Sulfur-Iodine (S-I) Cycle

Bunsen reaction (T = 20 – 100 °C):

$$SO_{2(g)} + I_{2(s)} + 2H_2O_{(l)} \leftrightarrow 2HI_{(aq)} + H_2SO_{4(aq)}$$

O_2 generation (T = 800 – 900 °C):

$$H_2SO_{4(g)} \leftrightarrow H_2O_{(g)} + SO_{2(g)} + \frac{1}{2}O_{2(g)}$$

H_2 generation (T = 400 – 550 °C):

$$2HI_{(g)} \leftrightarrow H_{2(g)} + I_{2(g)}$$

Net reaction:

$$H_2O_{(l)} \rightarrow H_{2(g)} + O_{2(g)}$$

The Vanadium-Chloride (V-Cl) Cycle

VCl₃ decomposition (T = 535°C):

$$2VCl_{3(s)} \rightarrow Cl_{2(g)} + 2VCl_{2(s)}$$

O_2 generation (T = 100 °C):

$$H_2O_{(g)} + Cl_{2(g)} \rightarrow \frac{1}{2}O_{2(g)} + 2HCl_{(g)}$$

H_2 generation (T = 300 °C):

$$2HCl_{(g)} + 2VCl_{2(s)} \rightarrow H_{2(g)} + 2VCl_{3(s)}$$

Net reaction:

$$H_2O_{(g)} \rightarrow H_{2(g)} + O_{2(g)}$$

FIGURE 3.2 An illustration of the water splitting process into H_2 and O_2 using a nonstoichiometric metal oxide redox pair and concentrated solar energy in two steps. (a) In the two-step process, redox materials are heated (at high temperatures) to create O_2 vacancies. Subsequently, steam is contacted with the partially-reduced redox material at lower temperatures, which scavenges for O_2 and releases H_2; (b) The sulfur-iodine (S-I) thermochemical cycle; and (c) the vanadium–chloride (V-Cl) cycle, showing the process of H_2 evolution.

In general, thermochemical cycles have the potential to achieve high efficiencies, such as 40%–56% for Zn/ZnO cycles, 39%–45% for Fe_3O_4/FeO cycles, 35%–46% for hybrid sulfur cycles, and 40%–60% for S-I and Cu-Cl cycles [47]. These efficiencies are comparable to current steam methane reforming methods [4, 47]. Another advantage of these cycles is that the water decomposition process can be separated from the solar thermal energy capturing and processing facilities. Therefore, the design and maintenance of these facilities can be carried out independently.

After conducting a thorough investigation, Brown and colleagues concluded that the most efficient thermochemical process for generating H_2 is the V-Cl cycle [48]. This conclusion was also reached by Safari and Dincer [39] in their extensive review. Up to 42.5% overall efficiency has been reported for V-Cl cycles [49], and this has been similarly reported in the work of Behzade and colleagues [50], where they incorporated the V-Cl thermochemical cycle in a hybrid methane-solar fuel multigeneration system (Figure 3.2). They obtained an exergy efficiency of 42.27% at optimal operation, which resulted in the production of 0.0266 kg/s of H_2 and 3,854 kW of electricity. However, the reported exergy efficiency is significantly lower than that reported (77%) elsewhere [39] for the V-Cl cycle, and may be a result of the thermodynamic conditions adopted. The assessment performed by Safari and Dincer [39] also concluded that the S-I and HyS cycles are the most promising in terms of global warming potential, whereas, when examining the integration of a thermochemical cycle with waste nuclear heat, the hybrid Cu-Cl cycle has the greatest potential [39].

3.5 NANOMATERIALS FOR THERMOCHEMICAL WATER SPLITTING

During multiple TCWS cycles, the utilized redox materials (Figure 3.2) undergo thermal stresses, which cause particle sintering and grain growth. These result in specific surface area and pore volume reduction and consequently a change in the characteristic packed bed volume of the redox material. The ultimate impact of these occurrences is a reduction in the volume of H_2 generation. The synthesis of nanomaterials capable of mitigating these challenges may involve mixing ceramic nanoparticles (such as ZrO_2), which possess high thermal stability, with sol-gel derived Ni-ferrite nanoparticles, which is an exceptional redox material for high-volume H_2 generation [51]. However, ordinary vortex mixing may induce a heterogeneous distribution of the particles in the final mixture. Thus, a synthesis method (e.g. solvothermal reaction, atomic layer deposition or polymeric precursor coating) that utilizes a core material with high H_2 generation capacity, encapsulated by a shell nanoparticle of excellent thermal stability yields a more homogeneous material capable of maintaining the H_2 generation volumes over multiple TCWS cycles. Amar et al. [52] synthesized coreshell $NiFe_2O_4$/Y_2O_3 nanoparticles for the attainment of a more stable H_2 generation volume via thermochemical water splitting at a temperature range of 900°C–1100°C. Their approach proved successful compared to the use of only $NiFe_2O_4$ nanoparticles and a powdered mixture of $NiFe_2O_4$/Y_2O_3 nanoparticles, which both revealed a continuously decreasing H_2 generation profile over multiple thermochemical cycles.

Figure 3.3 illustrates the transient performance of different nanoparticles in relation to the obtained H_2 generation stability. It can be observed that a relatively stable

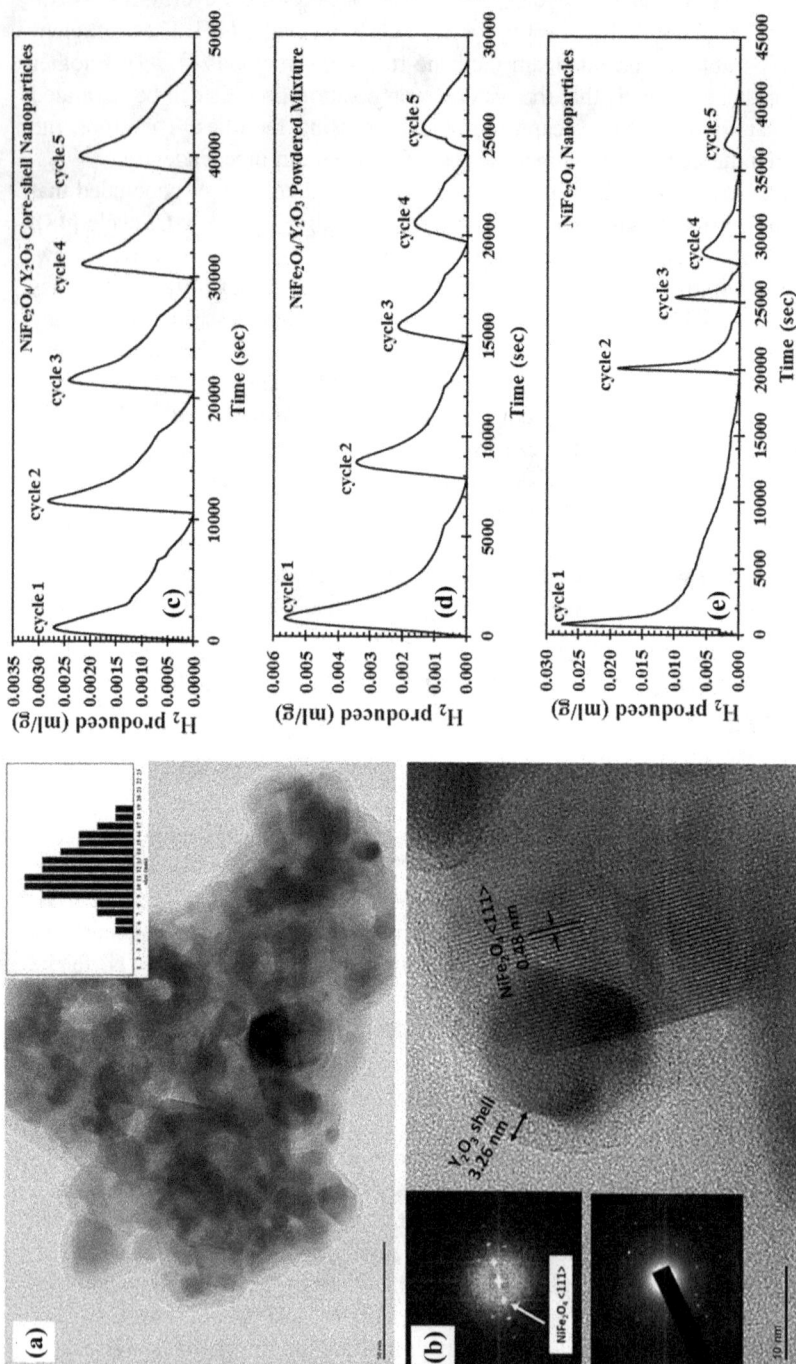

FIGURE 3.3 Transmission electron microscopy (TEM) image of agglomerated NiFe$_2$O$_4$/Y$_2$O$_3$ core-shell nanoparticles showing: (a) the nanoparticle size distribution, and (b) the core-shell morphology. Transient H$_2$ generation profiles, recorded during five thermochemical cycles for three different nanomaterials: (c) NiFe$_2$O$_4$/Y$_2$O$_3$ core-shell nanoparticles, (d) NiFe$_2$O$_4$/Y$_2$O$_3$ powdered mixture, and (e) NiFe$_2$O$_4$ nanoparticles. The water-splitting and regeneration steps were carried out at temperatures of 900°C and 1100°C, respectively [52].

profile exists for the core-shell nanoparticles, whereas, the powdered mixture and the $NiFe_2O_4$ nanoparticles exhibited a continuously decreasing generation volume of H_2. Nonetheless, the average H_2 volume generation was higher for the powdered mixture than that of the core-shell structure – a consequence of mass transfer limitations (the Y_2O_3 shell acting as a diffusional barrier for the TCWS reactions).

In the Fe_3O_4/FeO cycle, the redox activity of ferrites is usually hindered due to the sluggish diffusion of Fe^{2+} ions [53]. To surpass this limitation, two popular methods can be applied: The nanostructuring of iron oxides or the addition of extra supports like yttrium-stabilized cubic zirconia (YSZ), ZrO_2, and calcium-stabilized cubic zirconia (CaSZ) [54]. Nanostructuring has been accomplished by depositing $CoFe_2O_4$ and $NiFe_2O_4$ through atomic layer deposition on a high surface area of Al_2O_3 to enhance reaction kinetics [55]. Furthermore, the slow rate of H_2 evolution is a common limitation of low-temperature multi-step cycles. For example, the Mn_3O_4/MnO cycle, despite being void of corrosive products (a prevalent challenge of most cycles), suffers from this problem of poor H_2 evolution kinetics [56]. However, the use of the key reactants in the form of nanoparticles has been highlighted as a reliable means for improving the rate of H_2 production – a consequence of the increased surface area provided [53]. Dey et al. [57] demonstrated an overall increase of 1.5–2 times in the H_2 generation rate through the use of ball-milled materials compared to bulk samples of the material (Mn_3O_4 and Na_2CO_3) in the Mn_3O_4/MnO cycle). Additionally, the application of nanoparticles further reduced the temperature requirement.

Bhaskarwar and co-workers have extensively studied the S-I cycle using a variety of nanomaterials. Singhania and Bhaskarwar [58] investigated the application of metal-doped (La, Nd, Pr) ceria nanoparticles (5–10 nm) for the catalytic decomposition of hydrogen iodide (HI) in the S-I TCWS cycle for H_2 production. It was observed that this doping strategy improved the thermal stability (up to 35 h), specific surface area and O_2 vacancy concentration; CeO_2-La exhibited the best performance. Furthermore, the high catalytic activity and stability of dispersed palladium nanocatalysts, supported on carbon nanotubes (2–20 nm), for HI decomposition in the S-I thermochemical cycle has also been demonstrated by the same authors [59]. It was realized that after 100 h of operation, no changes in the catalyst's particle size, structure, specific surface area and elemental analysis were observed. An equilibrium conversion of 23.7% at 550°C was achieved in this TCWS cycle, demonstrating the suitability of the synthesized Pd(3%)/CNT catalyst. Singhania et al. [60] further examined the efficacy of bimetallic Ni-Pt nanoparticles (≤10 nm) supported on zirconia, ceria, γ-alumina and activated carbon) for H_2 production via the S-I TCWS cycle. Of all supports, activated carbon performed best in terms of catalytic activity, achieving up to 23.9% HI conversion at 550°C. This was followed by γ-alumina, zirconia, and ceria. The Ni(2.5%)-Pt(2.5%)/AC catalyst was stable for up to 100 h, in the HI decomposition step of the S-I TCWS cycle. More recently, the popular photocatalyst (TiO_2) has been shown to be suitable for thermochemical H_2 production via the water-splitting S-I cycle (with 50% efficiency) [61]. The synthesized nanoparticle revealed superior catalytic activity compared to a commercial TiO_2 nanocatalyst. A reasonable stability of up to 6 h was demonstrated for continuous HI decomposition; albeit this is significantly lower than the stabilities reported by other nanocatalysts developed by the same authors [58–60]. Nonetheless, with this success, the authors recommended further exploration of this catalyst for low-cost thermochemical H_2 production.

Nanofluids (a weak solution of nanoparticles) also find relevance as heat transfer enhancers in the operation of TCWS cycles; this is a result of their excellent thermal conductivity. Zhou et al. [62] utilized Al_2O_3 and CuO nanoparticles in a heat exchanger to enhance the heat transfer to the V-Cl thermochemical cycle for increased H_2 generation. It was realized that Al_2O_3 nanoparticles increased the heat transfer efficiency by 18.5% compared to the base fluid and by 8.3% compared to CuO particles. In the event that solar thermal energy is transported over a distance from a solar tower to a H_2 production cycle through a heat transfer fluid, it is necessary to limit heat losses to below 30% to be competitive with water electrolysis and steam methane reforming H_2 production processes [4]. Thus, the use of nanofluids has great potential in this regard.

3.6   CHALLENGES OF NANOMATERIAL-BASED THERMOCHEMICAL PROCESSES FOR RENEWABLE H₂ PRODUCTION

It is anticipated that nanomaterials will continue to play a crucial role in the generation of renewable biofuels through the modification of the reaction mechanisms, as prolific catalysts. Depending on the particular thermochemical process, the required quantity of a suitable nanocatalyst differs considerably. However, it is evident that the scale-up of laboratory TCWS processes will require a commensurate scale-up in the production routes of these nanomaterials. The bulk synthesis of these nanomaterials through conventional chemical routes (chemical vapor deposition, sol-gel techniques, precipitation, chemical etching) requires several harmful chemicals at each step, and these tend to produce intermediates and byproducts, which are harmful and toxic to the environment, and to living organisms. Conversely, physical synthesis methods (sputtering, ball milling, physical vapor deposition, electro-deposition) can be expensive and too time-consuming for large-scale applications.

Although green synthesis routes (from microorganisms – including bacteria, fungi and algae, plant extracts, animal tissues and biomass) have been proposed, the yield of produced nanoparticles tends to be significantly lower than the conventional synthesis methods. Microorganism-based synthesis is also typically plagued by the high cost of culture preparation and maintenance, coupled with the risk of infection to the workers. Furthermore, the scarcity of information on the specific bioactive groups crucial to the synthesis and their mechanisms of action is a major limitation of biological synthesis routes. Srivastava et al. [63] have reported that the formation of polydispersed nanomaterials is also a critical limitation of green synthesis routes for nanocatalysts. More technological advancements are required for the efficient control of morphological characteristics (size and shape) of nanomaterials synthesized by green/biological routes; this will facilitate their seamless integration into thermochemical H_2 production processes.

The separation of a catalyst from pyrolysis residues after biomass conversion is also a key challenge notable with thermochemical H_2 production from biomass. This causes carbon deposition and eventual catalyst deactivation; however, this happens at

TABLE 3.2
Operational Challenges of some TWCS Cycles

Cycle	Key Operational Challenges
2-step (ZnO/Zn)	High material cost for high-temperature solar concentrators.
3-step (S-I & HyS)	Risk of corrosion and high temperature for H_2SO_4 decomposition; difficult separation of HI from H_2SO_4.
4-step (V-Cl)	Poorly understood thermodynamics and slow chlorination kinetics. Potential recombination of produced gases.
4-step (Mg-Cl)	Rapid $MgCl_2$ hydrates formation, which could constitute a solids-handling problem.
4-step (Fe-Cl)	High-temperature requirement of $FeCl_3$ decomposition and its dimerization due to H_2 bond formation in the presence of water. Heavy metal processing also poses a challenge.
4-step (Cu-Cl)	Involves the use of corrosive fluids, and typically requires solid handling between steps.

a reduced rate with nanocatalysts. Although a promising method to mitigate catalyst deactivation, the direct impregnation of nanocatalysts onto the biomass matrix is plagued by several factors such as the limited catalyst loading (due to the low specific surface area of biomass), heterogenous catalyst distribution (which may cause uneven reaction rates and poor conversion), and difficult catalyst separation and recovery. Additionally, impurities present in the biomass can interfere with the catalytic activity, leading to reduced H_2 production.

Table 3.2 highlights some of the specific challenges of different TCWS cycles. The problems of separation of intermediate products from the main products and corrosiveness, particularly in the S-I cycle are also bound to affect nanomaterials utilized in this process. Thus, these challenges, although typical of the main cycle itself also raise concerns, due to the potential interference with the activity and stability of an adopted nanocatalyst. Since high temperatures are typically required for the TCWS process, the selection of materials for solar concentrating equipment (troughs, heliostats, reflection mirrors, lenses, parabolic dishes) and the appropriate working fluids (molten salts, thermal oils and nanofluids) is a crucial engineering challenge. Although thermal oils are typically utilized in solar devices, they may decompose at the temperature conditions of these cycles, and this causes the release of toxic fumes [4]. Molten salts have the drawback of having higher melting points compared to thermal oils and gases, which imposes limitations on their ability to transfer and store heat efficiently. These high temperatures associated with thermochemical conversion processes may also induce sintering, morphological variation, heterogeneity, reduced selectivity, and deactivation of the utilized nanocatalysts. These can be addressed via the development of porous carbon coatings and metallic coatings, as this increases the thermal stability of the catalyst. Mesoporous silica-coated nanoparticles are stable at temperatures as high as 727°C as demonstrated by Yue et al. [64].

Despite the previously described advancements, TCWS cycles are not yet considered to be cost-competitive compared to conventional methods such as SMR. Thus, catalytic improvements of existing cycles and the discoveries of new nanocatalysts should be accompanied with techno-economic analyses and life-cycle assessments to ascertain their potential for large-scale synthesis and utilization. Furthermore, there are very few large-scale experimental campaigns of theoretically viable TCWS cycles using nanocatalysts. The pursuit of new developments in this regard will be vital to achieving commercial H_2 production via these cycles.

3.7   CONCLUSIONS

In this chapter, the relevance of nanomaterials for thermochemical H_2 production has been extensively outlined. A key challenge for which nanomaterials find direct relevance is the reduction of the high-temperature requirements for achieving reasonably high H_2 yields. To overcome the limitations of heat and mass transfer, thermochemical reactors that incorporate nanocatalysts must be cleverly designed with high-durability materials while taking into consideration, the catalyst's stability over numerous production cycles. Embedding metallic nanocatalysts directly into the matrix of different biomass sources is a crucial development that can be incorporated into the biomass pretreatment processes to facilitate a more compact operation while reducing the potential for tar formation. While two-step TCWS cycles are not as thermodynamically favorable as three- and four-step cycles, it will be advantageous to investigate alternative oxide materials with lower temperature requirements. Conducting a thermodynamic analysis of different oxides would aid the rapid screening and identification of potential materials. Although a nonrenewable route, thermochemical H_2 production from fossil fuels, will further benefit from novel and sustainable nanomaterial synthesis technologies where their size and shape can be readily controlled. The transition from chemical-based nanomaterial synthesis to biological synthesis is a reasonable step towards reducing (albeit indirectly) the carbon footprint of SMR processes. Furthermore, the integration of H_2 production units into systems where energy is wasted is of high importance. Power-to-gas (PTG) technologies (the utilization of excessive electricity from an intermittent renewable energy system for H_2 production) and hybrid TCWS cycles will go a long way towards decarbonizing the current energy sector, amidst long-term plans for fully sustainable energy production.

REFERENCES

1. Council, H. Hydrogen scaling up: A sustainable pathway for the global energy transition. 2017. https://www.h2knowledgecentre.com/content/policypaper1201
2. Martín, J.G. The future of hydrogen: Seizing todays opportunities. *Econ. Ind.* 2022, *424*, 183–184.
3. Herron, J.A.; Kim, J.; Upadhye, A.A.; Huber, G.W.; Maravelias, C.T. A general framework for the assessment of solar fuel technologies. *Energy Environ. Sci.* 2015, *8*, 126–157.
4. Wang, Z.; Roberts, R.R.; Naterer, G.F.; Gabriel, K.S. Comparison of thermochemical, electrolytic, photoelectrolytic and photochemical solar-to-hydrogen production technologies. *Int. J. Hydrogen Energy* 2012, *37*, 16287–16301.
5. Mohsin, M.; Ishaq, T.; Bhatti, I.A.; Jilani, A.; Melaibari, A.A.; Abu-Hamdeh, N.H. Semiconductor nanomaterial photocatalysts for water-splitting hydrogen production: The Holy Grail of converting solar energy to fuel. *Nanomaterials* 2023, *13*, 546.

6. Ishaq, T.; Yousaf, M.; Bhatti, I.A.; Batool, A.; Asghar, M.A.; Mohsin, M.; Ahmad, M. A perspective on possible amendments in semiconductors for enhanced photocatalytic hydrogen generation by water splitting. *Int. J. Hydrogen Energy* 2021, *46*, 39036–39057.

7. Brun, K.; Allison, T.C. *Machinery and Energy Systems for the Hydrogen Economy*; Elsevier, 2022; ISBN 0323903940.

8. Qayyum, H., Cheema, I.I., Abdullah, M., Amin, M., Khan, I.A., Lee, E.J. and Lee, K.H., 2023. Modeling of one dimensional heterogeneous catalytic chemical looping steam methane reforming in an adiabatic packed bed reactor. *Front. Chem. 11*, 1295455. https://doi.org/10.3389/fchem.2023.1295455

9. Epelle, E.I.; Desongu, K.S.; Obande, W.; Adeleke, A.A.; Ikubanni, P.P.; Okolie, J.A.; Gunes, B. A comprehensive review of hydrogen production and storage: A focus on the role of nanomaterials. *Int. J. Hydrogen Energy* 2022. *47*, 20398–20431.

10. Kim, H.; Lee, Y.-H.; Lee, H.; Seo, J.-C.; Lee, K. Effect of Mg contents on catalytic activity and coke formation of mesoporous Ni/Mg-aluminate spinel catalyst for steam methane reforming. *Catalysts* 2020, *10*, 828.

11. Lai, G.-H.; Lak, J.H.; Tsai, D.-H. Hydrogen production via low-temperature steam–methane reforming using Ni–CeO2–Al2O3 hybrid nanoparticle clusters as catalysts. *ACS Appl. Energy Mater.* 2019, *2*, 7963–7971.

12. Vinum, M.G.; Almind, M.R.; Engbæk, J.S.; Vendelbo, S.B.; Hansen, M.F.; Frandsen, C.; Bendix, J.; Mortensen, P.M. Dual-function Cobalt–Nickel nanoparticles tailored for high-temperature induction-heated steam Methane reforming. *Angew. Chemie* 2018, *130*, 10729–10733.

13. Ligthart, D.; Van Santen, R.A.; Hensen, E.J.M. Influence of particle size on the activity and stability in steam methane reforming of supported Rh nanoparticles. *J. Catal.* 2011, *280*, 206–220.

14. De Maron, J.; Mafessanti, R.; Gramazio, P.; Orfei, E.; Fasolini, A.; Basile, F. H2 production by Methane Oxy-reforming: Effect of catalyst pretreatment on the properties and activity of Rh-Ce0. 5Zr0. 5O2 synthetized by microemulsion. *Nanomaterials* 2022, *13*, 53.

15. Velazquez Abad, A.; Dodds, P.E. Production of hydrogen. *Encyclopedia of Sustainable Technologies* 2017. 293–304. https://www.sciencedirect.com/science/article/pii/B9780124095489101174?via%3Dihub

16. Ghahraloud, H.; Farsi, M.; Rahimpour, M.R. Hydrogen production through thermal decomposition of hydrogen sulfide: Modification of the sulfur recovery unit to produce ultrapure hydrogen. *Ind. Eng. Chem. Res.* 2018, *57*, 14114–14123.

17. Karimi, F.; Mazaheri, D.; Saei Moghaddam, M.; Matei Moghaddam, A.; Sanati, A.L.; Orooji, Y. Solid-state fermentation as an alternative technology for cost-effective production of bioethanol as useful renewable energy: A review. *Biomass Convers. Biorefinery* 2021, 1–17. https://doi.org/10.1007/s13399-021-01875-2

18. Karimi-Maleh, H.; Rajendran, S.; Vasseghian, Y.; Dragoi, E.-N. Advanced integrated nanocatalytic routes for converting biomass to biofuels: A comprehensive review. *Fuel* 2021, 122762.

19. Jafarian, S.; Tavasoli, A.; Karimi, A.; Norouzi, O. Steam reforming of bagasse to hydrogen and synthesis gas using ruthenium promoted NiFe/γAl2O3nano-catalysts. *Int. J. Hydrogen Energy* 2017, *42*, 5505–5512.

20. Nanda, S.; Reddy, S.N.; Dalai, A.K.; Kozinski, J.A. Subcritical and supercritical water gasification of lignocellulosic biomass impregnated with nickel nanocatalyst for hydrogen production. *Int. J. Hydrogen Energy* 2016, *41*, 4907–4921.

21. Norouzi, O.; Safari, F.; Jafarian, S.; Tavasoli, A.; Karimi, A. Hydrothermal gasification performance of Enteromorpha intestinalis as an algal biomass for hydrogen-rich gas production using Ru promoted Fe–Ni/γ-Al2O3 nanocatalysts. *Energy Convers. Manag.* 2017, *141*, 63–71.

22. Barati, M.; Babatabar, M.; Tavasoli, A.; Dalai, A.K.; Das, U. Hydrogen production via supercritical water gasification of bagasse using unpromoted and zinc promoted Ru/γ-Al$_2$O$_3$ nanocatalysts. *Fuel Process. Technol.* 2014, *123*, 140–148.

23. Rashidi, M.; Tavasoli, A. Hydrogen rich gas production via supercritical water gasification of sugarcane bagasse using unpromoted and copper promoted Ni/CNT nanocatalysts. *J. Supercrit. Fluids* 2015, *98*, 111–118.

24. Chen, F.; Wu, C.; Dong, L.; Jin, F.; Williams, P.T.; Huang, J. Catalytic steam reforming of volatiles released via pyrolysis of wood sawdust for hydrogen-rich gas production on Fe–Zn/Al$_2$O$_3$ nanocatalysts. *Fuel* 2015, *158*, 999–1005.

25. Tursun, Y.; Xu, S.; Abulikemu, A.; Dilinuer, T. Biomass gasification for hydrogen rich gas in a decoupled triple bed gasifier with olivine and NiO/olivine. *Bioresour. Technol.* 2019, *272*, 241–248.

26. Li, J.; Yan, R.; Xiao, B.; Liang, D.T.; Du, L. Development of nano-NiO/Al$_2$O$_3$ catalyst to be used for tar removal in biomass gasification. *Environ. Sci. Technol.* 2008, *42*, 6224–6229.

27. Collard, F.-X.; Blin, J.; Bensakhria, A.; Valette, J. Influence of impregnated metal on the pyrolysis conversion of biomass constituents. *J. Anal. Appl. Pyrolysis* 2012, *95*, 213–226.

28. Shen, Y.; Zhao, P.; Shao, Q.; Ma, D.; Takahashi, F.; Yoshikawa, K. In-situ catalytic conversion of tar using rice husk char-supported nickel-iron catalysts for biomass pyrolysis/gasification. *Appl. Catal. B Environ.* 2014, *152*, 140–151.

29. Islam, M.W. A review of dolomite catalyst for biomass gasification tar removal. *Fuel* 2020, *267*, 117095.

30. Cao, C.; Zhang, Y.; Cao, W.; Jin, H.; Guo, L.; Huo, Z. Transition metal oxides as catalysts for hydrogen production from supercritical water gasification of glucose. *Catal. Letters* 2017, *147*, 828–836.

31. Cao, C.; Xie, Y.; Mao, L.; Wei, W.; Shi, J.; Jin, H. Hydrogen production from supercritical water gasification of soda black liquor with various metal oxides. *Renew. Energy* 2020, *157*, 24–32.

32. Bion, N.; Duprez, D.; Epron, F. Design of nanocatalysts for green hydrogen production from bioethanol. *ChemSusChem* 2012, *5*, 76–84.

33. Zhang, B.; Zhang, L.; Yang, Z.; He, Z. An experiment study of biomass steam gasification over NiO/Dolomite for hydrogen-rich gas production. *Int. J. Hydrogen Energy* 2017, *42*, 76–85.

34. Arregi, A.; Amutio, M.; Lopez, G.; Bilbao, J.; Olazar, M. Evaluation of thermochemical routes for hydrogen production from biomass: A review. *Energy Convers. Manag.* 2018, *165*, 696–719.

35. Dong, L.; Wu, C.; Ling, H.; Shi, J.; Williams, P.T.; Huang, J. Promoting hydrogen production and minimizing catalyst deactivation from the pyrolysis-catalytic steam reforming of biomass on nanosized NiZnAlOx catalysts. *Fuel* 2017, *188*, 610–620.

36. Tavasoli, A.; Barati, M.; Karimi, A. Conversion of sugarcane bagasse to gaseous and liquid fuels in near-critical water media using K$_2$O promoted Cu/γ-Al$_2$O$_3$–MgO nanocatalysts. *Biomass and Bioenergy* 2015, *80*, 63–72.

37. Gao, W.; Farahani, M.R.; Rezaei, M.; Hosamani, S.M.; Jamil, M.K.; Imran, M.; Baig, A.Q. Experimental study of steam-gasification of municipal solid wastes (MSW) using Ni–Cu/γ-Al$_2$O$_3$ nano catalysts. *Energy Sources, Part A Recover. Util. Environ. Eff.* 2017, *39*, 693–697.

38. Adnan, M.A.; Adamu, S.; Muraza, O.; Hossain, M.M. Fluidizable NiO–Fe$_2$O$_3$/SiO$_2$–γAl$_2$O$_3$ for tar (toluene) conversion in biomass gasification. *Process Saf. Environ. Prot.* 2018, *116*, 754–762.

39. Safari, F.; Dincer, I. A review and comparative evaluation of thermochemical water splitting cycles for hydrogen production. *Energy Convers. Manag.* 2020, *205*, 112182.

40. Steinfeld, A. Solar hydrogen production via a two-step water-splitting thermochemical cycle based on Zn/ZnO redox reactions. *Int. J. Hydrogen Energy* 2002, *27*, 611–619.

41. Zhang, P.; Chen, S.Z.; Wang, L.J.; Yao, T.Y.; Xu, J.M. Study on a lab-scale hydrogen production by closed cycle thermo-chemical iodine–sulfur process. *Int. J. Hydrogen Energy* 2010, *35*, 10166–10172.

42. Anzieu, P.; Carles, P.; Le Duigou, A.; Vitart, X.; Lemort, F. The sulphur–iodine and other thermochemical process studies at CEA. *Int. J. Nucl. Hydrog. Prod. Appl.* 2006, *1*, 144–153.

43. Ishaq, H.; Dincer, I.; Naterer, G.F. Industrial heat recovery from a steel furnace for the cogeneration of electricity and hydrogen with the copper-chlorine cycle. *Energy Convers. Manag.* 2018, *171*, 384–397.

44. Lewis, M.A.; Masin, J.G. The evaluation of alternative thermochemical cycles–Part II: The down-selection process. *Int. J. Hydrogen Energy* 2009, *34*, 4125–4135.

45. Lewis, M.A.; Ferrandon, M.S.; Tatterson, D.F.; Mathias, P. Evaluation of alternative thermochemical cycles–Part III further development of the Cu–Cl cycle. *Int. J. Hydrogen Energy* 2009, *34*, 4136–4145.

46. Bilgen, E. Solar hydrogen production by hybrid thermochemical processes. *Sol. Energy* 1988, *41*, 199–206.

47. Roeb, M.; Thomey, D.; Graf, D.; de Oliveira, L.; Sattler, C.; Poitou, S.; Pra, F.; Tochon, P.; Brevet, A.; Roux, G. HycycleS–A project on solar and nuclear hydrogen production by sulphur-based thermochemical cycles. In Proceedings of the Proc. 18th World Hydrogen Energy Conf. WHEC; 2010; Vol. 18.

48. Brown, L.C.; Besenbruch, G.E.; Lentsch, R.D.; Schultz, K.R.; Funk, J.F.; Pickard, P.S.; Marshall, A.C.; Showalter, S.K. *High efficiency generation of hydrogen fuels using nuclear power*; General Atomics, San Diego, CA, 2003. 1–343.

49. Chehade, G.; Alrawahi, N.; Yuzer, B.; Dincer, I. A photoelectrochemical system for hydrogen and chlorine production from industrial waste acids. *Sci. Total Environ.* 2020, *712*, 136358.

50. Behzadi, A.; Gholamian, E.; Alirahmi, S.M.; Nourozi, B.; Sadrizadeh, S. A comparative evaluation of alternative optimization strategies for a novel heliostat-driven hydrogen production/injection system coupled with a vanadium chlorine cycle. *Energy Convers. Manag.* 2022, *267*, 115878.

51. Bhosale, R.R.; Shende, R.V.; Puszynski, J.A. Sol-gel derived $NiFe_2O_4$ modified with ZrO_2 for hydrogen generation from solar thermochemical water-splitting reaction. *MRS Online Proc. Libr.* 2012, 1387, mrsf11-1387. https://doi.org/10.1557/opl.2012.772

52. Amar, V.S.; Puszynski, J.A.; Shende, R. V H2 generation from thermochemical water-splitting using yttria stabilized $NiFe_2O_4$ core-shell nanoparticles. *J. Renew. Sustain. Energy* 2015, *7*, 23113.

53. Rao, C.N.R.; Dey, S. Solar thermochemical splitting of water to generate hydrogen. *Proc. Natl. Acad. Sci.* 2017, *114*, 13385–13393.

54. Gokon, N.; Murayama, H.; Umeda, J.; Hatamachi, T.; Kodama, T. Monoclinic zirconia-supported Fe_3O_4 for the two-step water-splitting thermochemical cycle at high thermal reduction temperatures of 1400–1600°C. *Int. J. Hydrogen Energy* 2009, *34*, 1208–1217.

55. Arifin, D.; Aston, V.J.; Liang, X.; McDaniel, A.H.; Weimer, A.W. $CoFe_2O_4$ on a porous Al_2O_3 nanostructure for solar thermochemical CO_2 splitting. *Energy Environ. Sci.* 2012, *5*, 9438–9443.

56. Xu, B.; Bhawe, Y.; Davis, M.E. Low-temperature, manganese oxide-based, thermochemical water splitting cycle. *Proc. Natl. Acad. Sci.* 2012, *109*, 9260–9264.

57. Dey, S.; Rajesh, S.; Rao, C.N.R. Significant reduction in the operating temperature of the Mn (II)/Mn (III) oxide-based thermochemical water splitting cycle brought about by the use of nanoparticles. *J. Mater. Chem. A* 2016, *4*, 16830–16833.

58. Singhania, A.; Bhaskarwar, A.N. Effect of rare earth (RE–La, Pr, Nd) metal-doped ceria nanoparticles on catalytic hydrogen iodide decomposition for hydrogen production. *Int. J. Hydrogen Energy* 2018, *43*, 4818–4825.

59. Singhania, A.; Bhaskarwar, A.N. Catalytic performance of carbon nanotubes supported palladium catalyst for hydrogen production from hydrogen iodide decomposition in thermochemical sulfur iodine cycle. *Renew. Energy* 2018, *127*, 509–513.

60. Singhania, A.; Krishnan, V. V.; Bhaskarwar, A.N.; Bhargava, B.; Parvatalu, D.; Banerjee, S. Catalytic performance of bimetallic Ni-Pt nanoparticles supported on activated carbon, gamma-alumina, zirconia, and ceria for hydrogen production in sulfur-iodine thermochemical cycle. *Int. J. Hydrogen Energy* 2016, *41*, 10538–10546.

61. Singhania, A.; Bhaskarwar, A.N. TiO_2 as a catalyst for hydrogen production from hydrogen-iodide in thermo-chemical water-splitting sulfur-iodine cycle. *Fuel* 2018, *221*, 393–398.

62. Zhou, J.; Ali, M.A.; Hai, T.; Sharma, K.; Aziz, K.H.; Alyousuf, F.Q.A.; Almoalimi, K.T.; Almojil, S.F.; Almohana, A.I.; Alali, A.F. Enhanced hydrogen generation in a combined hybrid cycle using aluminum and cooper oxide nanomaterial based on biomass and vanadium chloride cycle: Optimization based on deep learning techniques and Environmental appraisal. *Int. J. Hydrogen Energy* 2022. *147*, https://www.sciencedirect.com/science/article/pii/S1364032121004949?via%3Dihub

63. Srivastava, M.; Srivastava, N.; Saeed, M.; Mishra, P.K.; Saeed, A.; Gupta, V.K.; Malhotra, B.D. Bioinspired synthesis of iron-based nanomaterials for application in biofuels production: A new in-sight. *Renew. Sustain. Energy Rev.* 2021, *147*, 111206.

64. Yue, L.; Li, J.; Chen, C.; Fu, X.; Xia, X.; Hou, J.; Xiao, C.; Chen, X.; Zhao, L.; Ran, G. Thermal-stable Pd@ mesoporous silica core-shell nanocatalysts for dry reforming of methane with good coke-resistant performance. *Fuel* 2018, *218*, 335–341.

4 Biological Hydrogen Production
The Role and Potential of Nanomaterials

Okon-Akan Omolabake Abiodun,
Olugbenga Akande, Chukwuma C. Ogbaga,
Adekunle A. Adeleke, Peter Ikubanni,
Toheeb Jimoh, Jude A. Okolie, and
Oluwaseun Iyadunni Oluwasogo

4.1 INTRODUCTION

Hydrogen has the potential to provide solutions to pressing issues related to the increasing global energy demand. Notably, it offers economically practical, financially beneficial, socially favorable, and energetically efficient means to address concerns like global warming, climate change and environmental pollution [1]. Recent research emphasizes the importance of initiating and expediting the transition from traditional energy systems to sustainable alternatives [2]. This energy transition is crucial in tackling global warming and other challenges associated with conventional energy sources, with hydrogen serving as a foundational pillar [3]. Hydrogen systems can fulfill significant roles during this energy shift, encompassing cleaner transportation, reduced environmental impacts, enhanced sustainability, improved efficiency, and more [4].

Integrating highly robust energy systems into cohesive frameworks allows for the generation of a diverse range of valuable commodities with minimal losses. The widespread adoption and market penetration of renewable energy sources can be achieved through the availability of hydrogen [5]. Hydrogen serves as a valuable complement to electrical storage for intermittent renewables like solar and wind power. Projections suggest that by 2050, hydrogen could meet 18% of the global energy demand, resulting in a reduction of 6 Gt of CO_2 emissions annually and the creation of 30 million jobs [4]. Furthermore, research indicates that hydrogen has the potential to power a significant portion of the transportation industry by 2050, with estimates suggesting it could fuel over 400 million automobiles, 15–20 million trucks, and approximately 5 million buses, accounting for 20%–25% of the sector [6]. It is anticipated that future hydrogen energy systems will be more efficient, delivering substantial economic and environmental benefits [6]. Notably in the literature, hydrogen is considered a leading option for addressing global warming and fostering long-term economic growth [7].

While hydrogen holds great promise, there are challenges associated with its production [8]. Steam reforming of natural gas (SMR) is a widely used method for hydrogen production; however, it faces challenges due to its heavy reliance on fossil fuels and the environmental issues arising from greenhouse gas (GHG) emissions [9]. Therefore, there is an urgent need to develop alternative production methods that are cost-effective and environmentally friendly. Figure 4.1 illustrates the diverse range of thermochemical and biological methods that have been studied for hydrogen production. Among these methods, biological hydrogen production shows promise due to several advantageous factors: It offers improved energy conversion efficiency, lower temperature requirements, and greater flexibility and scalability [10].

These advantages make biological hydrogen production an attractive option for sustainable hydrogen generation. Biological systems offer a wide array of microorganisms and metabolic pathways that contribute to hydrogen production [11]. This diversity enables researchers to explore and optimize various biological systems, leading to enhanced process efficiency and adaptability. Biological hydrogen production pathways include anaerobic digestion (AD), direct and indirect bio-photolysis, photo fermentation, and dark fermentation [11]. However, these biological processes often exhibit lower hydrogen yields compared to theoretical values and require biomass pretreatment and specific cultured microorganisms [12].

To address these challenges, the integration of nanoparticles has been proposed [8]. Nanoparticles play a crucial role in advancing biological hydrogen production as they can enhance biomass pretreatment, improve enzymatic hydrolysis, and facilitate enzyme production [8]. Considering the significance of nanomaterials in this context, this chapter aims to discuss their role in advancing biological hydrogen production. The initial section provides an overview of different biological hydrogen production methods, highlighting their respective advantages and limitations. Subsequently, the chapter comprehensively examines the applications of nanomaterials at various stages of the biological hydrogen production process. By exploring the synergistic potential of nanomaterials and biological systems, this chapter seeks to contribute to the development of more efficient and sustainable methods for hydrogen production.

4.1.1 Overview of Biological Hydrogen Production Methods

Biological hydrogen production methods involve the utilization of microorganisms, bacteria, or enzymes to convert biogenic waste into hydrogen. These methods encompass various approaches such as photolysis, photo fermentation, dark fermentation, anaerobic digestion, and CO gas fermentation [8]. In bio-photolysis, microalgae and cyanobacteria harness solar energy to generate hydrogen through the conversion of water and CO_2. This process takes advantage of their photosynthetic capabilities [13]. Photo fermentation, on the other hand, relies on purple non-sulfur (PNS) bacteria. These bacteria utilize different carbon sources and convert them into hydrogen using light energy. Dark fermentation involves anaerobic bacteria that produce hydrogen from organic matter derived from food or agricultural waste, as well as wastewater [13].

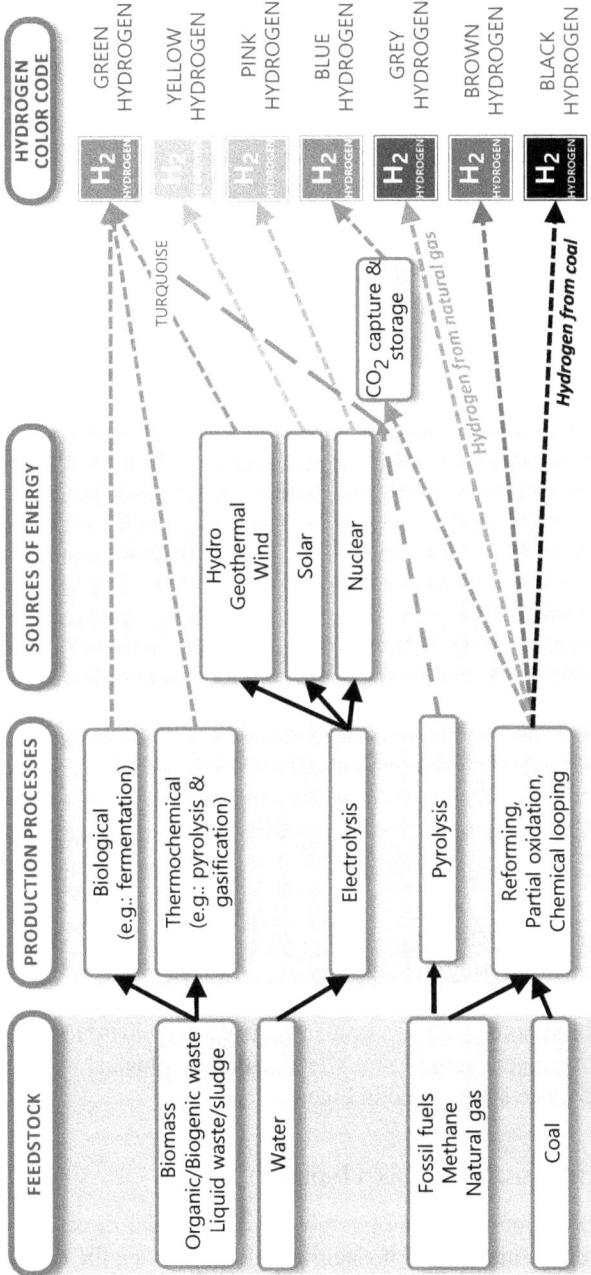

FIGURE 4.1 An overview of hydrogen production methods and the associated color codes [8].

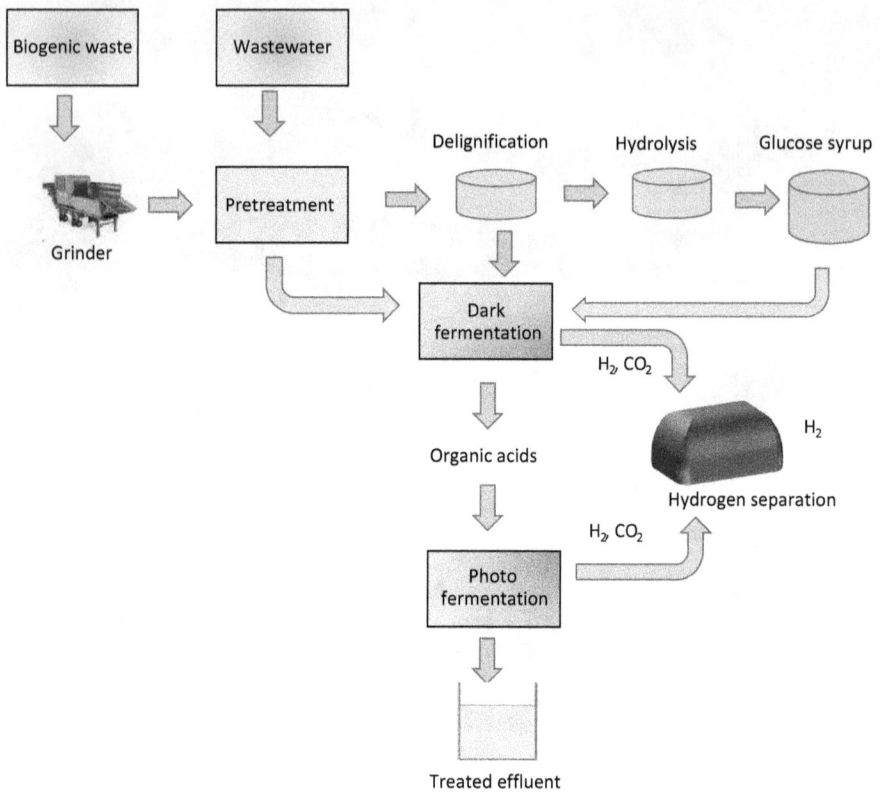

FIGURE 4.2   A schematic of biological hydrogen production process.

In CO gas-fermentation, hydrogen is produced through the bioconversion of carbon monoxide and water molecules via the water-gas shift reaction. This process occurs under anaerobic conditions, facilitated by photosynthetic bacteria [14]. These diverse biological hydrogen production methods utilize different microorganisms and metabolic pathways, each offering unique advantages and applications. By harnessing the potential of biogenic waste, solar energy, and anaerobic processes, biological hydrogen production holds promise as a sustainable and environmentally friendly approach to hydrogen generation [14]. A schematic of biological hydrogen production from organic waste is presented in Figure 4.2.

4.1.2   BIO-PHOTOLYSIS (DIRECT AND INDIRECT)

There are two subcategories of bio-photolysis, indirect and direct bio-photolysis. Indirect bio-photolysis makes use of electrons released during the catabolization of endogenous substrates, while direct bio-photolysis makes use of electrons released during the water splitting in photosystem II.

Biological organisms can undergo processes like bio-photolysis, solar-powered hydrolysis, or the separation of hydrogen and oxygen from water molecules [15].

Autotrophic organisms, including cyanobacteria and microalgae, can perform photo-synthesis, using carbon dioxide from the air and water from the soil, along with solar radiation, to produce carbohydrate biomass. This two-step process, known as indirect bio-photolysis, involves the initial generation of carbohydrates during photosynthe-sis, followed by their fermentation in the dark to produce hydrogen.

Indirect bio-photolysis involves the release of electrons from organic compounds such as starch and glycogen, which are stored within the organisms [16]. The first stage encompasses photosynthesis and carbohydrate accumulation, while the second stage entails fermenting the stored carbohydrates to generate hydrogen. To ensure successful hydrogen evolution, which is more delicate than oxygen evolution, these two phases must be carried out separately [16].

Direct bio-photolysis occurs when photoautotrophic organisms transform water molecules to hydrogen with the aid of hydrogenase enzymes as catalysts under anaerobic conditions and light energy [14]. The overall process is explained by Equations 4.1 and 4.2.

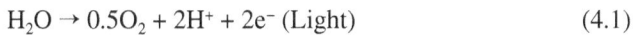

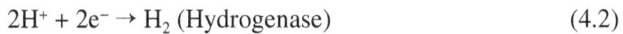

$$H_2O \rightarrow 0.5O_2 + 2H^+ + 2e^- \text{ (Light)} \tag{4.1}$$

$$2H^+ + 2e^- \rightarrow H_2 \text{ (Hydrogenase)} \tag{4.2}$$

In another study conducted by [9], it is explained that algae and cyanobacteria have the potential to be utilized in a process called direct bio-photolysis to produce hydrogen gas. During this process, a series of reactions occur, and it involves the col-laboration of photosystems (Photosystem II) and PSI (Photosystem I), which are typically found in plant-type photosynthesis. These photosystems play a crucial role in mediating the overall process of hydrogen generation. The pathway for these reac-tions is presented in Figure 4.3. It should be noted that the electron transport chain

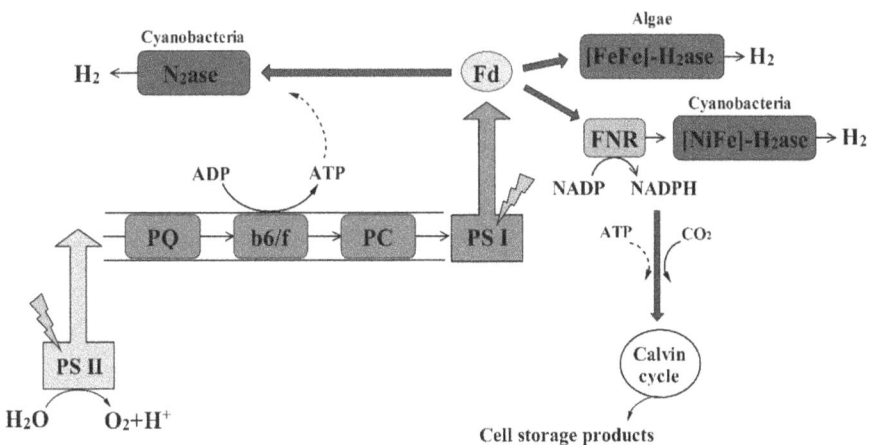

FIGURE 4.3 Pathways for the direct bio-photolysis of green algae with cyanobacteria for hydrogen production.

(Adapted from [14] with permission from Elsevier.)

involved in photosynthesis includes important components such as plastoquinone (PQ), cytochromes b6 and f, and plastocyanin (PC). These components play essential roles in facilitating the flow of electrons during the process. Additionally, FNR, which stands for ferredoxin-NADP+ oxidoreductase, is an enzyme involved in transferring electrons from ferredoxin to NADP+ to generate NADPH, a crucial molecule in photosynthetic reactions.

Photosystem II makes use of solar energy to produce hydrogen and oxygen as byproducts of water splitting during the bio-photolysis process. By using nitrogenase or hydrogenase enzymes to reduce protons, hydrogen is subsequently generated. PSII and PSI are part of the electron transport chain which involves a water-splitting process and electron flux through the thylakoid membrane. A ferredoxin molecule is reduced in the process, and adenosine triphosphate (ATP) is produced [16]. These numerous metabolic processes generate hydrogen.

4.1.3 DARK AND PHOTO FERMENTATION

Dark fermentation, a process performed by obligate and facultative anaerobes, is a productive and efficient method of hydrogen production. It occurs in the absence of light and oxygen, and during this process, anaerobic bacteria break down carbohydrate-rich substrates, such as lignocellulosic biomass, industrial effluent, crop residues high in sugar content, and municipal solid waste (MSW), to produce hydrogen [17]. The success of dark fermentation is influenced by factors such as biomass pretreatment, microorganism selection, and sugar availability in the substrate. However, the accumulation of metabolites, like volatile fatty acids, can limit the yield of hydrogen production in dark fermentation [18]. On the other hand, photo fermentation occurs when photosynthetic bacteria undergo fermentation in the presence of light. This process is distinct from dark fermentation as it relies on the deactivation of nitrogenase due to light energy [18, 19]. Photosynthetic bacteria utilize organic substrates, such as acetate, butyrate, and lactate, to produce hydrogen and carbon dioxide through a series of biochemical events [20]. Compared to dark fermentation, photo fermentation can potentially achieve higher hydrogen yields. However, the efficiency of hydrogen production through photo fermentation is lower, and it requires larger reactor sizes compared to dark fermentation [3]. Hybrid systems which combine the advantages of both dark fermentation and photo fermentation have been developed [3]. They also address the challenges associated with the byproducts of dark fermentation while maximizing hydrogen production [3]. Hybrid systems offer the potential for increased efficiency and sustainability in hydrogen generation.

In dark fermentation, a wide range of materials can be used as substrates, including lignocellulosic biomass, food scraps, MSW, sewage, and glycerol. Lignocellulosic biomass, in particular, holds promise as a cost-effective and abundant source of carbohydrates for biohydrogen production [18]. The origin of the substrate, the source of the inoculum, and various operating parameters, such as pH, temperature, and hydraulic retention time, influence biohydrogen production and yield [18].

Dark fermentation faces challenges, including limited hydrogen production caused by the accumulation of volatile fatty acids in the fermentation medium [21]. To address these challenges, researchers have investigated the simultaneous operation of

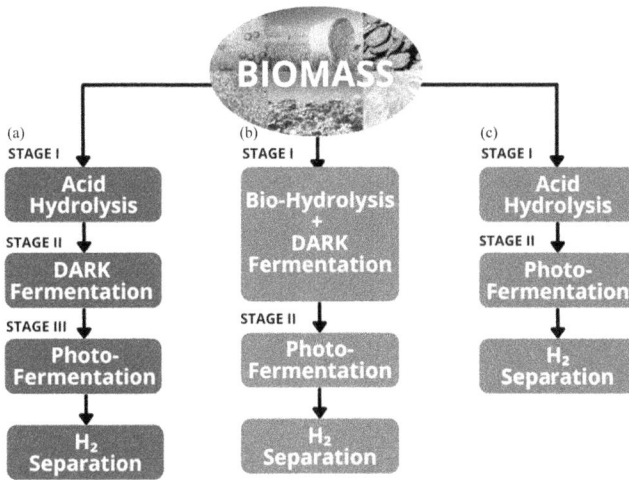

FIGURE 4.4   An overview of sequential dark and photo fermentation processes for hydrogen production. (a) Three-step process: Acid hydrolysis followed by dark and photo fermentations. (b) Two-step process: Bio-hydrolysis with simultaneous dark fermentation followed by photo fermentation. (c) Two-step process: Acid hydrolysis followed by photo fermentation.

dark and photo fermentation processes as a potential solution [21]. Previous studies have shown that a sequential combination of dark and photo fermentation reactions can lead to an increase in the maximum theoretical yield of hydrogen [22]. Figure 4.4 depicts three different configurations of sequential dark and photo fermentation processes used for biohydrogen production.

The technical challenges associated with each biological method for hydrogen production are summarized in Figure 4.5. Based on these challenges, nanomaterials are applied at various stages to improve hydrogen yield. The next section discusses the role of nanomaterials in improving hydrogen yield.

4.2   NANOMATERIALS FOR ENHANCING BIOLOGICAL HYDROGEN PRODUCTION

4.2.1   NANOMATERIALS FOR DARK FERMENTATION OF HYDROGEN PRODUCTION

Dark fermentation, as mentioned earlier is one of the modern ecofriendly techniques for producing biofuels (biodiesel, biomethane, biohydrogen), with biohydrogen being the cleanest of them all [23]. Dark fermentation has been found to produce an excellent yield of biohydrogen on a laboratory scale, but, despite being light-independent, it has been reported to produce biohydrogen of low quantity and quality on a larger scale. This is undoubtedly due to the formation of byproducts which include alcohols, soluble metabolites, and volatile organic acids [24]. Hence, there is a need to find new sources and metabolic pathways to enhance this technique's ability to produce a high yield of biohydrogen on a larger scale. Nanomaterials have been developed recently to improve the synthesis of biohydrogen via dark fermentation

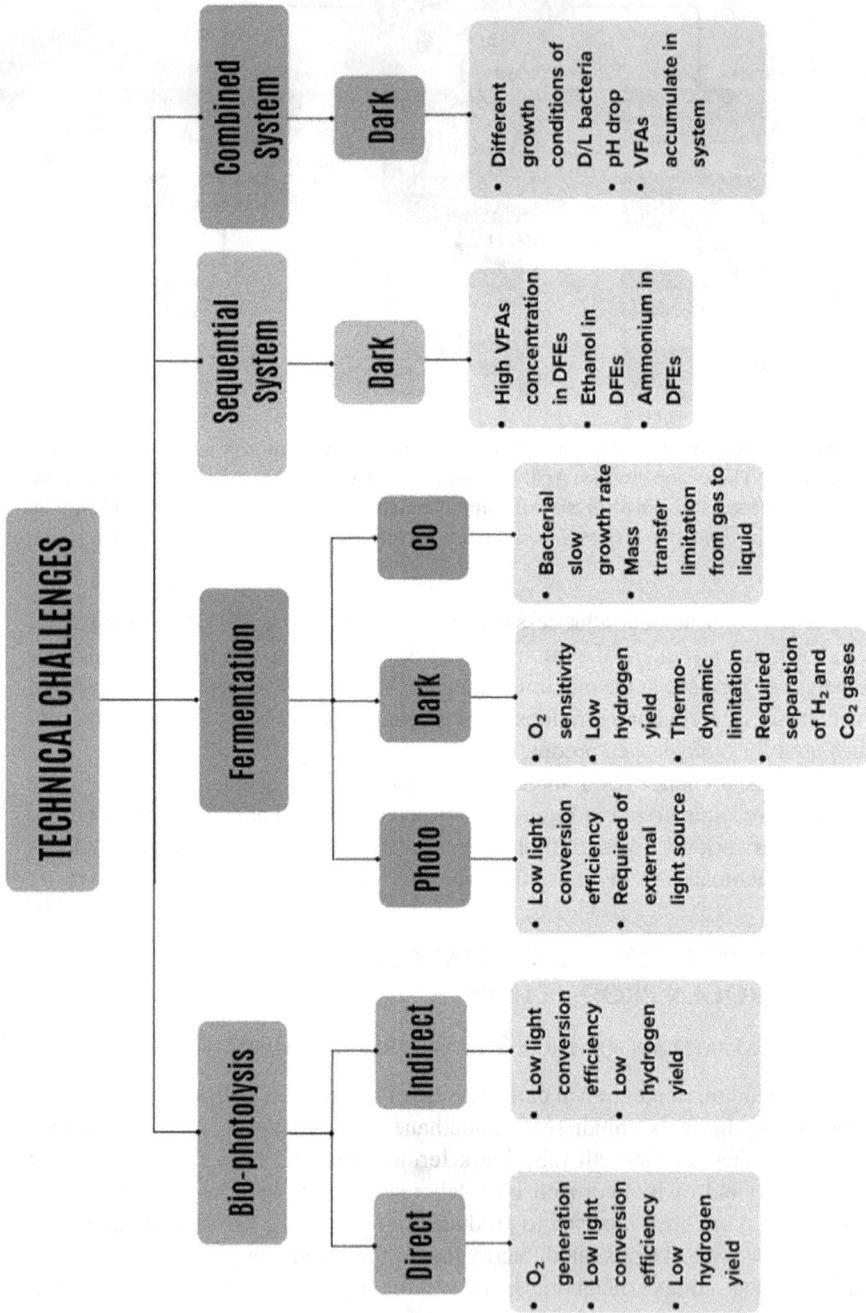

FIGURE 4.5   An overview of the challenges associated with biological hydrogen production.

owing to their distinctive physicochemical properties (excellent catalytic activity, good specificity, and larger surface area) [23].

Hydrogenase enzymes are not only responsible for the production of biological hydrogen via dark fermentation but also for the degradation of organic substrate anaerobically and in the absence of light [23, 25]. Hydrogenase enzymes, based on the metals in the active sites, are of two types namely: Ni-Fe hydrogenase and Fe-Fe hydrogenase. Nanomaterials have been found to improve biohydrogen yield and productivity by specifically targeting hydrogenase enzymes. For instance, [26] reported a 32.6% hydrogen production increment in the fermentation of sucrose when hematite (Fe_2O_3) was added to the process. More so, [27], in their study revealed that biohydrogen yield increased by 22.7% when nickel nanoparticles were added to the fermentation of glucose. More so, nanomaterials, especially magnetic ones such as iron and nickel, serve as cofactors for these hydrogenase enzymes, thereby enhancing the efficiency of the process by modifying the enzyme's activity and stability. [28] reported a very high biohydrogen yield of 150 mL/g at a temperature of 310 K and pH of 7 while studying the effects of nanomaterials ($Fe°$ and $Ni°$) on dark fermentative biohydrogen production in a batch reactor.

Nanomaterials catalyze the formation of biohydrogen through their small size and large surface area which offer a greater number of sites for enzymatic interaction with the substrate. They have a significant impact on a microorganism's metabolic activity by promoting effective electron transfer to acceptors, increasing productivity and biohydrogen yield [24]. Additionally, nanomaterials' quantum size and large surface area boost electron absorption and improve the rate of electron transfer between the nanomaterials and enzymes, hence enhancing the production of hydrogen. For example, due to gold nanoparticles' high surface-area-to-volume ratio, which had a stimulating influence on the production of biohydrogen, 5 nm gold nanoparticles increased substrate utilization and biohydrogen yield by 56% and 46%, respectively [29]. [30] did a comparative study to investigate the effect of palladium nanomaterials and palladium chloride ($PdCl_2$) on dark fermentative biohydrogen production. Their results revealed that the addition of the former to *Enterobacter cloacae* resulted in about a 2% increment yield of hydrogen production, while the chlorides caused a 5% decrease in the yield of biohydrogen produced.

It is important to mention that a nanomaterial's mechanism in dark fermentative biohydrogen production is dependent on microorganisms and concentration even if it enhances biohydrogen yield. Several studies on nanomaterials have concluded that the optimal concentration for hematite, Fe_2O_3 for biohydrogen production is 200 mg/L, and any concentration greater than this will result in oxidative damage to the bacteria, causing cell death, which will directly impact the bacteria's yield and growth rate in the reaction mixture [23]. Hence, it is essential to use nanomaterials at their threshold value to prevent enzyme inhibition and cell death.

4.2.2 NANOMATERIALS FOR ENHANCING PHOTO FERMENTATIVE HYDROGEN PRODUCTION

The photo fermentative technique is a light-dependent process and can also be used for biohydrogen production. Unlike the dark fermentative technique, the nitrogenase enzyme is responsible for mediating the photo fermentative technique. The introduction of nanomaterials in photo fermentative techniques increases the biohydrogen yield by boosting the nitrogenase enzyme's activity while decreasing the activity of the hydrogenase [25]. Nitrogenase-containing microorganisms can be immobilized on nanomaterials, which boosts their thermo- and pH- stability and enables them to flourish in a wide range of temperatures and pH [31]. Immobilization also improves the enzyme's reusability and storage capacity, which ensures the process's environmental friendliness and cost-effectiveness [30]. Nanomaterials enhance photo fermentative bacteria's growth and metabolism, which increases the amount of biomass involved in the production of biohydrogen.

Nanomaterials can also significantly improve the yield of biohydrogen produced via the photo fermentative technique by acting as an electron carrier in the electron transport channel, which ensures that the electrons are rapidly transferred to their corresponding electron acceptor and also by binding the nitrogenase enzyme's active site which ensures that the process is catalyzed [25].

Enzyme's light conversion efficiency and light intensity are some of the factors that can affect the photo fermentative technique. Nanomaterials, especially the photo-catalytic ones (ZnO and TiO_2) can improve the photo fermentative bacteria's photo-conversion efficiency and provide the necessary energy required for the synthesis of biohydrogen [32]. These photo-catalytic nanomaterials have also been found to promote photo-catalysis and surface adsorption because of their quantum size effects, enhanced reactive surface area, and increased number of active sites. More so, they have the potential to boost biohydrogen's yield and productivity by ensuring that electrons are rapidly transferred to the enzyme system (nitrogenase) [25, 33]. For example, [34] in their study investigating the effect of photo-catalytic TiO_2 nanomaterials on *Rhodobacter sphaeroides* photo fermentative biohydrogen production revealed that under optimum condition of 305 K, the biohydrogen production, substrate conversion efficiency, and production rate increased by 1.7-fold, 63.3%, and 1.5-fold respectively.

4.2.3 NANOMATERIALS FOR ENHANCING BIOMASS PRETREATMENT

Even though biomass has been recognized as one of the raw materials with huge potential for renewable energy, its complex structure calls for pretreatment so that its potential can be harnessed fully. The conventional pretreatment techniques (chemical, biological, and physical) that were used for the treatment come with certain drawbacks that include, but are not limited to, environmental pollution, expensive processing cost, degradation of products, and low processing efficiency [29]. Hence, there is a need to search for a novel, cost-effective, and environmentally friendly technique that can ensure that the biomass' potential is fully used.

Nanomaterials have been touted as a better alternative as they have recently shown significant progress in the biomass-based industry by reducing the drawbacks of conventional pretreatment techniques [29]. Nanomaterials have special qualities; among them is an excellent surface area to volume ratio, which is needed for effective processing of biomass [35]. Nanomaterials also offer great potential for enzyme recovery and reusability in scaling up biofuel production. They have also proved useful in immobilizing hydrolyzing enzymes for biomass.

Nanomaterials have been reported to use immobilization, adsorption, and absorption techniques in treating biomass [29]. Furthermore, it has been stressed that immobilizing enzymes on nanomaterials will enhance their functional capacity when used in the best possible concentrations and conditions. Nanomaterial-based photocatalysts and magnetic nanoparticles have been found to offer a cutting-edge technique and efficient means of treating biomass because of their fast reactions, multiple reusability, and being easily separated [36].

The use of nanomaterial-based photocatalysts has been acknowledged as one of the important methods for the pretreatment of biomass and, hence, has been extensively researched. It has been attempted to pretreat lignin in order to selectively cleave the C–C bond using a nanomaterial-based photocatalyst for a variety of value additions [37]. For instance, [32] pretreated lignin in the presence of UV radiation using a TiO_2–ZnO nanocomposite that was prepared via ex-situ technique. This was done by exposing the lignin to the radiation for four hours while 4% of the nanocomposite was used to generate the targeted oxidation. Their results revealed that the methane yield increased by 39% in comparison to the control system and more than 80% of the lignin was degraded which attests to the effectiveness of the composite. More so, [38] subjected lignin to degradation through photocatalysis in a batch-recirculated photoreactor using a TiO_2/polystyrene nanocomposite prepared by solvent casting technique. They then revealed that more than 90% of the lignin was degraded, and more significantly, the TiO_2/polystyrene nanocomposite's reusability, which was examined after five cycles, even then showed good stability and conspicuous photocatalysis efficiency.

Therefore, it can be inferred from the aforementioned research that the pretreatment of biomass with nanomaterial-based photocatalysts offers a promising means of producing biofuels in a renewable manner. However, the reaction has to be controlled as it is crucial, otherwise, undesired byproducts will be produced, and this will further hamper the production of targeted biofuels [39].

Acid-functionalized magnetic nanocatalysts have huge potential to eliminate the need for acidic chemical regents and the negative effects of inhibitors during the pretreatment process. More so, their high surface-to-volume ratio offers strong catalytic activity and, hence, aids a high rate of hydrolysis which translates to a high yield of the targeted product [40]. Another benefit of using these types of nanomaterial is that they are simple to separate when the reaction is complete for additional reuse, which can significantly lower the cost of the entire bioprocess [29]. [41] did a comparative study to investigate the effectiveness of an acid (H_2SO_4) and acid-functionalized magnetic nanocatalysts (Fe_3O_4-Si-AS and Fe_3O_4-Si-BCOOH) in the pretreatment of sugarcane straw. They found that the pretreatment mediated by the nanocatalysts was more effective as it produced higher sugar than that of the H_2SO_4. [42] used hematite (Fe_3O_4) nanoparticles to pretreat rice straws while attempting to enhance the production of biogas. It was discovered that in comparison to the control system, the methane and biogas yield increased by 33% and 37% respectively when 100 ppm of the hematite nanoparticle was used.

Carbon nanotubes (CNTs), on the other hand, have also received a lot of attention for biomass pretreatment due to their exceptional design, electrical strength, excellent thermal characteristics, and outstanding tensile durability [29]. They are made of graphite sheets that are arranged in a hollow and round shape. They can be single-walled or multi-walled depending on the number of layers. The latter appear to be more effective at immobilizing cellulase due to their affordability, enhancement of electrical properties, high physical and chemical stability, and low toxicity [35]. Numerous researchers have looked into the use of CNTs for immobilizing cellulase. In the review by [40], recovered cellulase from *Aspergillus niger* has been immobilized on multi-walled carbon nanotubes. The authors reported that the immobilized cellulase maintained 85% of its enzyme activity and was able to hydrolyze the cellulose numerous times at a substantial pH and constant temperature.

4.2.4 NANOMATERIALS FOR ENHANCING BIOMASS HYDROLYSIS

Hydrolysis of biomass is an important stage in biofuel production as it is the stage where the overall efficiency is determined. This could be done via the chemical method and enzyme catalysis [40]. The former uses acid catalysts such as H_2SO_4 and HCl for hydrolysis and it is associated with limitations such as lower efficiency, poor specificity, and production of harmful inhibitors [44]. The latter has emerged as the better choice because the enzymes' excellent specificity and promising catalytic abilities allow for environmentally friendly and efficient degradation of biomass [40]. The expensive operational cost caused by the enzymes' poor stability and reusability has stimulated the search for a promising approach.

Immobilization of enzymes on nanomaterials is a novel technique for the hydrolysis of biomass, and it is reported that it enhances enzymes' catalytic activity while it also gives them resilience against environmental attacks [40]. The immobilized enzymes on nanomaterials are referred to as "nanobiocatalyst". Nanomaterials such as silica nanoparticles, carbon nanotubes, magnetic nanomaterials, oxide nanoparticles, and nickel particles have been employed in supporting the immobilization of enzymes. Nanomaterials, because of their high surface area and increased enzyme loading per unit mass can improve the efficiency of the immobilized enzymes [45].

Many researchers are currently interested in the immobilization of enzymes on magnetic nanoparticles because they offer many benefits including biocompatibility, high specific surface area, low toxicity, low resistance to mass transfer, and low stability. It has been shown that enzymes are more thermostable when immobilized on solid supports like magnetic nanoparticles [29]. [46] in their research used glutaraldehyde as a coupling molecule to immobilize cellulase on chitosan-coated magnetic nanoparticles covalently. Their results revealed that the immobilized enzymes still maintain about 80% of their capacity even after several repeated use.

Silica nanoparticles also proved useful in immobilizing lignocellulolytic enzymes like cellulase [40]. For instance, [47] did a comparative study to investigate the catalytic activity of free cellulase enzyme and cellulose immobilized on silica nanoparticles in simultaneous saccharification and fermentation (SSF) reactions for enhanced ethanol production from lignocellulosic materials. Their results revealed that the silica nanoparticles displayed greater affinity for the cellulase adsorption, and also yielded about 1.6 times more glucose than the free cellulose enzyme.

Nickel nanoparticles have been used as a catalyst in converting lignocellulosic materials to biofuels. In general, catalytic reactions demand high temperatures, which can be reduced by employing nickel nanoparticles [40]. When lignocellulosic material is being hydrolyzed, the stability of the enzyme is an extremely important stage. Thus, the use of nanoparticles can enhance the stability of cellulases [29]. Nickel-cobaltite nanoparticles were produced, and their effect on the formation of cellulases at different concentrations was assessed by [48]. The study showed that an indirect relationship exists between nanoparticle concentration and the cellulase produced, as 1 mM concentration gave the highest rate of cellulase production.

The carbon nanotubes (CNTs) also cannot be ignored as they have also received a lot of attention due to their exceptional design, excellent thermal character, and electrical strength [49]. After six hydrolysis cycles, the catalytic activity of the cellulase that was immobilized on functionalized multi-walled carbon nanotubes was found to be 52% [50].

Other nanoparticles such as oxide nanomaterials can also be used to immobilize cellulase in addition to the ones discussed above. For instance, [51] in their study revealed that *Aspergillus fumigatus*-derived MnO_2 nanoparticles are an excellent support for immobilizing cellulase because they make the cellulase more thermostable at 70°C and that even after five hydrolysis cycles, the immobilized cellulase can still function at 60% of its peak level.

4.2.5 NANOTECHNOLOGY DEVICES FOR BIOHYDROGEN PRODUCTION

Research has been tilting toward developing nanotechnological devices for biohydrogen production after the breakthrough of nanomaterial synthesis. Microfluidic systems seem to be breakthrough research owing to their efficient mass and heat transfer, uniform flow, ability to build durable and compact fuel processors, large surface-to-volume ratio, and excellent photon transfer. They can be scaled down to microbial electrolysis cells (MEC), microbial fuel cells (MFC) or microbial electrochemical cells (MXC) [52]. Numerous advantages of microfluidics can be exploited to successfully produce biological hydrogen at the microscale or even to optimize its yield at a large scale by gathering data on parameters like light irradiation or cell behavior [53]. To build a self-powered biohydrogen generator that will generate hydrogen at the rates of 28 and 46 ppm/h with urea-based and glucose substrates respectively, [54] combined a microfluidic microbial electrolysis cell (MEC) and a microfluidic microbial electrochemical cell (MXC) with non-pathogenic *E. coli* strain being employed as biocatalyst. Microfluidic lab-on-a-chip systems combine several biological test processes and can precisely regulate, monitor, and alter samples at the nano- to pico-liter dimensions, making them suitable for building a high throughput screening platform [53].

4.3 CONCLUSION

This chapter provides a comprehensive review of various methods for biological hydrogen production, including light-dependent and independent fermentation, as well as bio-photolysis. It outlines the unique characteristics and restrictions of each method, thereby setting the stage for future explorations into how nanomaterials can push forward biological hydrogen production. Metals like nickel, iron, and copper are essential in refining the conversion process from biomass to biohydrogen.

Iron and nickel operate like auxiliary substances at the active regions of nitrogenase and hydrogenase enzymes, thus improving the yield of biohydrogen production. Past research also suggests that incorporating magnetic nanoparticles can alter the heat resilience and pH of cellulase enzymes and minimize the initial slow phase of microbial growth, resulting in enhanced hydrolysis. Furthermore, nanomaterials could be harnessed to boost hydrogen production via light-independent and dependent fermentation.

REFERENCES

1. Kumar G, Shobana S, Nagarajan D, Lee DJ, Lee KS, Lin CY, et al. Biomass based hydrogen production by dark fermentation – recent trends and opportunities for greener processes. *Curr Opin Biotechnol*. 2018; 50:136–45.

2. Das D, Veziroğlu TN. Hydrogen production by biological processes: A survey of literature. *Int J Hydrogen Energy*. 2001; 26(1):13–28.

3. Zhang Q, Zhang Z. Biological hydrogen production from renewable resources by photofermentation. In: *Advances in Bioenergy*. Elsevier; 2018. p. 137–60.

4. Acar C, Dincer I. Review and evaluation of hydrogen production options for better environment. *J Clean Prod*. 2019; 218:835–49.

5. Singh V, Das D. Potential of hydrogen production from biomass. *Sci Eng Hydrog energy Technol*. 2019; 123–64. https://doi.org/10.1016/B978-0-12-814251-6.00003-4

6. Ni M, Leung DYC, Leung MKH, Sumathy K. An overview of hydrogen production from biomass. *Fuel Process Technol*. 2006; 87(5):461–72.

7. Okolie JA, Patra BR, Mukherjee A, Nanda S, Dalai AK, Kozinski JA. Futuristic applications of hydrogen in energy, biorefining, aerospace, pharmaceuticals and metallurgy. *Int J Hydrogen Energy*. 2021; 46(13):8885–905.

8. Epelle EI, Desongu KS, Obande W, Adeleke AA, Ikubanni PP, Okolie JA, et al. A comprehensive review of hydrogen production and storage: A focus on the role of nanomaterials. *Int J Hydrogen Energy*. 2022; 47:20398–20431.

9. Łukajtis R, Hołowacz I, Kucharska K, Glinka M, Rybarczyk P, Przyjazny A, et al. Hydrogen production from biomass using dark fermentation. *Renew Sustain Energy Rev*. 2018; 91:665–94.

10. Okolie JA, Epelle EI, Tabat ME, Orivri U, Amenaghawon AN, Okoye PU, et al. Waste biomass valorization for the production of biofuels and value-added products: A comprehensive review of thermochemical, biological and integrated processes. *Process Saf Environ Prot*. 2022; 159:323–44.

11. Ramprakash B, Lindblad P, Eaton-Rye JJ, Incharoensakdi A. Current strategies and future perspectives in biological hydrogen production: A review. *Renew Sustain Energy Rev*. 2022; 168:112773.

12. Moussa RN, Moussa N, Dionisi D. Hydrogen production from biomass and organic waste using dark fermentation: An analysis of literature data on the effect of operating parameters on process performance. *Processes*. 2022; 10(1):156.

13. Najafpour G, Younesi H, Mohamed AR. Effect of organic substrate on hydrogen production from synthesis gas using Rhodospirillum rubrum, in batch culture. *Biochem Eng J*. 2004; 21(2):123–30.

14. Akhlaghi N, Najafpour-Darzi G. A comprehensive review on biological hydrogen production. Int J Hydrogen Energy. 2020; 45(43):22492–512.

15. Kamran U, Park SJ. Chemically modified carbonaceous adsorbents for enhanced CO2 capture: A review. *J Clean Prod*. 2021; 290:125776.

16. Ahmed SF, Rafa N, Mofijur M, Badruddin IA, Inayat A, Ali MS, et al. Biohydrogen production from biomass sources: Metabolic pathways and economic analysis. *Front Energy Res*. 2021; 9:753878.
17. Kamran U, Park SJ. Hybrid biochar supported transition metal doped MnO2 composites: Efficient contenders for lithium adsorption and recovery from aqueous solutions. *Desalination*. 2022; 522:115387.
18. Singh N, Sarma S. Biological routes of hydrogen production: A critical assessment. In: *Handbook of Biofuels*. Elsevier; 2022. p. 419–34.
19. Ren N, Guo W, Liu B, Cao G, Ding J. Biological hydrogen production by dark fermentation: Challenges and prospects towards scaled-up production. *Curr Opin Biotechnol*. 2011; 22(3):365–70.
20. Mili M, Hashmi SAR, Ather M, Hada V, Markandeya N, Kamble S, et al. Novel lignin as natural-biodegradable binder for various sectors – A review. *J Appl Polym Sci*. 2022; 139(15):51951.
21. Argun H, Kargi F. Bio-hydrogen production by different operational modes of dark and photo-fermentation: An overview. *Int J Hydrogen Energy*. 2011; 36(13):7443–59.
22. Ozmihci S, Kargi F. Bio-hydrogen production by photo-fermentation of dark fermentation effluent with intermittent feeding and effluent removal. *Int J Hydrogen Energy*. 2010; 35(13):6674–80.
23. Kumar G, Mathimani T, Rene ER. ScienceDirect Application of nanotechnology in dark fermentation for enhanced biohydrogen production using inorganic nanoparticles. *Int J Hydrogen Energy*. 2019; 44(26):13106–13.
24. Patel SKS, Kalia J, Kalia V.C. Nanoparticles in Biological Hydrogen Production: An Overview. Indian *J Microbiol*. 2017; 58:8–18.
25. Srivastava N, Srivastava M, Mishra PK, Adnan M, Saeed M, Gupta VK, et al. Bioresource Technology Advances in nanomaterials induced biohydrogen production using waste biomass. *Bioresour Technol*. 2020; 307(February):123094.
26. Han H, Cui M, Wei L, Yang H, Shen J. Enhancement effect of hematite nanoparticles on fermentative hydrogen production. *Bioresour Technol*. 2011; 102(17):7903–9.
27. Mullai P, Yogeswari MK, Sridevi K. Optimisation and enhancement of biohydrogen production using nickel nanoparticles – A novel approach. *Bioresour Technol*. 2013; 141:212–9.
28. Taherdanak M, Zilouei H, Karimi K. Investigating the effects of iron and nickel nanoparticles on dark hydrogen fermentation from starch using central composite design. *Int J Hydrogen Energy*. 2015; 40(38):12956–63.
29. Chandel H, Kumar P, Chandel AK, Verma ML. Biotechnological advances in biomass pretreatment for bio - renewable production through nanotechnological intervention. *Biomass Convers Biorefinery*. 2024; 2024(0123456789).
30. Mohanraj S, Kodhaiyolii S, Rengasamy M, Pugalenthi V. Phytosynthesized iron oxide nanoparticles and ferrous iron on fermentative hydrogen production using Enterobacter cloacae: Evaluation and comparison of the effects. *Int J Hydrogen Energy*. 2014; 39(23):11920–9.
31. Liu B, Jin Y, Wang Z, Xing D, Ma C, Ding J, et al. Enhanced photo-fermentative hydrogen production of Rhodopseudomonas sp. nov. strain A7 by the addition of TiO2, ZnO and SiC nanoparticles. *Int J Hydrogen Energy*. 2017; 42(29):18279–87.
32. Chu YM, Javed HMA, Awais M, Khan MI, Shafqat S, Khan FS, et al. Photocatalytic pretreatment of commercial lignin using TiO2-zno nanocomposite-derived advanced oxidation processes for methane production synergy in lab scale continuous reactors. *Catalysts*. 2021; 11(1):54.

33. Wang Y, Xiao G, Wang S, Su H. Bioresource Technology Reports Application of nano-materials in dark or light-assisted fermentation for enhanced biohydrogen production: A mini-review. *Bioresour Technol Reports*. 2023; 21(September 2022):101295.

34. Pandey A, Gupta K, Pandey A. Effect of nanosized TiO_2 on photofermentation by Rhodobacter sphaeroides NMBL-02. *Biomass and Bioenergy*. 2015; 72:273–9.

35. Gaikwad S, Ingle AP, da Silva SS, Rai M. Immobilized nanoparticles-mediated enzymatic hydrolysis of cellulose for clean sugar production: A novel approach. *Curr Nanosci*. 2019; 15(3):296–303.

36. Sindhu R, Binod P, Pandey A. Biological pretreatment of lignocellulosic biomass–An overview. *Bioresour Technol*. 2016; 199:76–82.

37. Soo H Sen, Gazi S, Dokic M. Selective photocatalytic CC bond cleavage under ambient conditions with earth abundant vanadium complexes. In: Abstracts of Papers of the American Chemical Society. Amer Chemical Soc, Washington, DC; 2016.

38. Haghighi M, Rahmani F, Kariminejad F, Sene RA. Photodegradation of lignin from pulp and paper mill effluent using TiO2/PS composite under UV-LED radiation: Optimization, toxicity assessment and reusability study. *Process Saf Environ Prot*. 2019; 122:48–57.

39. Srivastava N, Singh R, Srivastava M, Mohammad A, Harakeh S, Pratap R, et al. Bioresource technology impact of nanomaterials on sustainable pretreatment of lignocellulosic biomass for biofuels production: An advanced approach. *Bioresour Technol*. 2023; 369(December 2022):128471.

40. Rai M, Ingle AP, Pandit R, Paralikar P, Biswas K, Silverio S. Emerging role of nano-biocatalysts in hydrolysis of lignocellulosic biomass leading to sustainable bioethanol production. *Catal Rev*. 2018; 00(00):1–26.

41. Ingle AP, Philippini RR, da Silva SS. Pretreatment of sugarcane bagasse using two different acid-functionalized magnetic nanoparticles: A novel approach for high sugar recovery. *Renew Energy*. 2020; 150:957–64.

42. Khalid MJ, Waqas A, Nawaz I. Synergistic effect of alkaline pretreatment and magnetite nanoparticle application on biogas production from rice straw. *Bioresour Technol*. 2019; 275:288–96.

43. Shaheen TI, Emam HE. Sono-chemical synthesis of cellulose nanocrystals from wood sawdust using acid hydrolysis. *Int J Biol Macromol*. 2018; 107:1599–606.

44. Singh N, Dhanya BS, Verma ML. Nano-immobilized biocatalysts and their potential biotechnological applications in bioenergy production. *Mater Sci Energy Technol*. 2020; 3:808–24.

45. Grewal J, Ahmad R, Khare SK. Development of cellulase-nanoconjugates with enhanced ionic liquid and thermal stability for in situ lignocellulose saccharification. *Bioresour Technol*. 2017; 242:236–43.

46. Sanchez-Ramirez J, Martinez-Hernandez JL, Segura-Ceniceros P, Lopez G, Saade H, Medina-Morales MA, et al. Cellulases immobilization on chitosan-coated magnetic nanoparticles: Application for Agave Atrovirens lignocellulosic biomass hydrolysis. *Bioprocess Biosyst Eng*. 2017; 40:9–22.

47. Lupoi JS, Smith EA. Evaluation of nanoparticle-immobilized cellulase for improved ethanol yield in simultaneous saccharification and fermentation reactions. *Biotechnol Bioeng*. 2011; 108(12):2835–43.

48. Srivastava N, Rawat R, Sharma R, Oberoi HS, Srivastava M, Singh J. Effect of nickel–cobaltite nanoparticles on production and thermostability of cellulases from newly isolated thermotolerant Aspergillus fumigatus NS (Class: Eurotiomycetes). *Appl Biochem Biotechnol*. 2014; 174:1092–103.

49. Sajeev E, Shekher S, Ogbaga CC, Desongu KS, Gunes B, Okolie JA. Application of nanoparticles in bioreactors to enhance mass transfer during syngas fermentation. *Encyclopedia*. 2023; 3(2):387–95.

50. Mubarak NM, Wong JR, Tan KW, Sahu JN, Abdullah EC, Jayakumar NS, et al. Immobilization of cellulase enzyme on functionalized multiwall carbon nanotubes. *J Mol Catal B Enzym*. 2014; 107:124–31.

51. Cherian E, Dharmendirakumar M, Baskar G. Immobilization of cellulase onto MnO_2 nanoparticles for bioethanol production by enhanced hydrolysis of agricultural waste. *Chinese J Catal*. 2015; 36(8):1223–9.

52. Ocheni SI, Ogbaga CC, Mohammed SSD, Mangse G. Generating Bioelectricity from Traditional Food Processing Wastewater Using an Inoculum of Return Activated Sewage Sludge. In: International Conference on Innovations and Interdisciplinary Solutions for Underserved Areas. Springer; 2022. p. 187–94.

53. Karthikeyan B, Velvizhi G. ScienceDirect A state-of-the-art on the application of nano-technology for enhanced biohydrogen production. *Int J Hydrogen Energy*. 2023; (xxxx).

54. Fadakar A, Mardanpour MM, Yaghmaei S. The coupled microfluidic microbial electrochemical cell as a self-powered biohydrogen generator. *J Power Sources*. 2020; 451:227817.

5 Nanomaterials for Electrolytic and Photolytic Hydrogen Production
Production Methods, Challenges, and Prospects

Winifred Obande, Emmanuel I. Epelle, and Jude A. Okolie

5.1 INTRODUCTION

The path to achieving carbon neutrality and establishing a sustainable energy future will depend greatly on the advancement of a robust clean hydrogen (H_2) economy. Goldman Sachs [1] asserts that H_2 is uniquely positioned to enable more effective decarbonization of the global economy. With its potential to reduce greenhouse gas emissions by approximately 15% and CO_2 emissions by around 20%, green H_2 has the capacity to play a major role in achieving sustainability goals. In fact, it is estimated that it could fulfill 18%–25% of the world's energy demands by 2050. Notwithstanding, if trends over the last few decades continue, meeting global net-zero ambitions may be challenging. Despite predictions up until the mid-2000s that the global H_2 economy would come to fruition by 2040, the development and deployment of the relevant technologies and infrastructure have not progressed as expected. The International Energy Agency (IEA) [2] projects a rise in global hydrogen demand, reaching 115 Mt by 2030, up from the peak demand of 94 Mt in 2021 [1]. However, as of 2021, clean H_2 production capacity – that is, from non-fossil fuel sources – had not attained sufficient maturity for practical utilization (<1 Mt in 2021). The primary obstacle hindering the widespread adoption of clean hydrogen across multiple sectors is the high cost of production.

As the world grapples with a pressing energy crisis, it has become imperative to create policies that reconcile the need for energy security with the urgency of tackling climate change. Hydrogen as an energy carrier, with its high energy density could serve as a vital tool in ensuring energy security by reducing our reliance on non-renewable fuels. This can be achieved by substituting fossil fuels with hydrogen in end-use applications and transitioning from fossil-based H_2 production to relying

DOI: 10.1201/9781003371007-5

on renewable energy sources. Much of the current H_2 supply comes from reforming natural gas, the extraction of biomass, and the gasification of coal, petroleum, coke, and heavy oil. However, to fully unlock the potential of hydrogen, innovative and sustainable production methods such as electrolytic and photocatalytic H_2 production must be optimized as clean and efficient pathways for large-scale hydrogen generation, while minimizing greenhouse gas emissions.

This chapter provides a comprehensive overview of electrolytic (Section 5.2) and photolytic (Section 5.3) hydrogen production, with emphasis on the roles of nanomaterials in driving these technologies forward. While various photolytic and electrolytic methods exist, the discourse presented herein, centers on water splitting using renewable energy via electrolytic (alkaline, solid oxide, and proton exchange membrane) and photolytic (photocatalytic and photoelectrochemical) techniques as depicted in Figure 5.1. We delve into the importance of nanomaterials as electrocatalysts for electrolysis, and also explore nanophotocatalysts for photocatalytic clean H_2 production. Finally, some challenges and future prospects associated with their implementation are highlighted.

5.2 ELECTROLYTIC HYDROGEN PRODUCTION

Water splitting currently accounts for less than 5% of the global H_2 supply, and there is great potential to advance processes like electrolysis to significantly increase its contribution in the future. A typical electrolytic water-splitting reaction can be divided into two half-reactions: hydrogen evolution reaction (HER) and oxygen evolution reaction (OER). The HER is the reduction half-reaction that occurs at the cathode. It involves the reduction of protons (H+) from the electrolyte to form hydrogen gas (H_2), represented by Equation 5.1. Similarly, OER is the oxidation half-reaction that occurs at the anode. It involves the oxidation of water molecules to release oxygen gas (O_2) and protons (H+) as depicted in Equation 5.2.

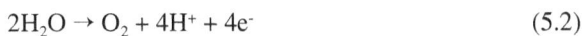

$$2H^+ + 2e^- \rightarrow H_2 \qquad (5.1)$$

$$2H_2O \rightarrow O_2 + 4H^+ + 4e^- \qquad (5.2)$$

The HER process begins with the Volmer reaction, a discharge process, characterized by the adsorption of a hydrogen atom unto the active site of the catalytic surface, and ensues as shown in Equation 5.3.

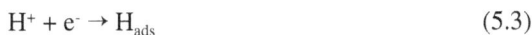

$$H^+ + e^- \rightarrow H_{ads} \qquad (5.3)$$

The adsorbed hydrogen is then desorbed either by an electrochemical (Heyrovsky) or chemical (Tafel) reaction, both of which are represented in Equations 5.4 and 5.5, respectively.

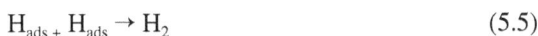

$$H_{ads +} H^+ + e^- \rightarrow H_2 \qquad (5.4)$$

$$H_{ads +} H_{ads} \rightarrow H_2 \qquad (5.5)$$

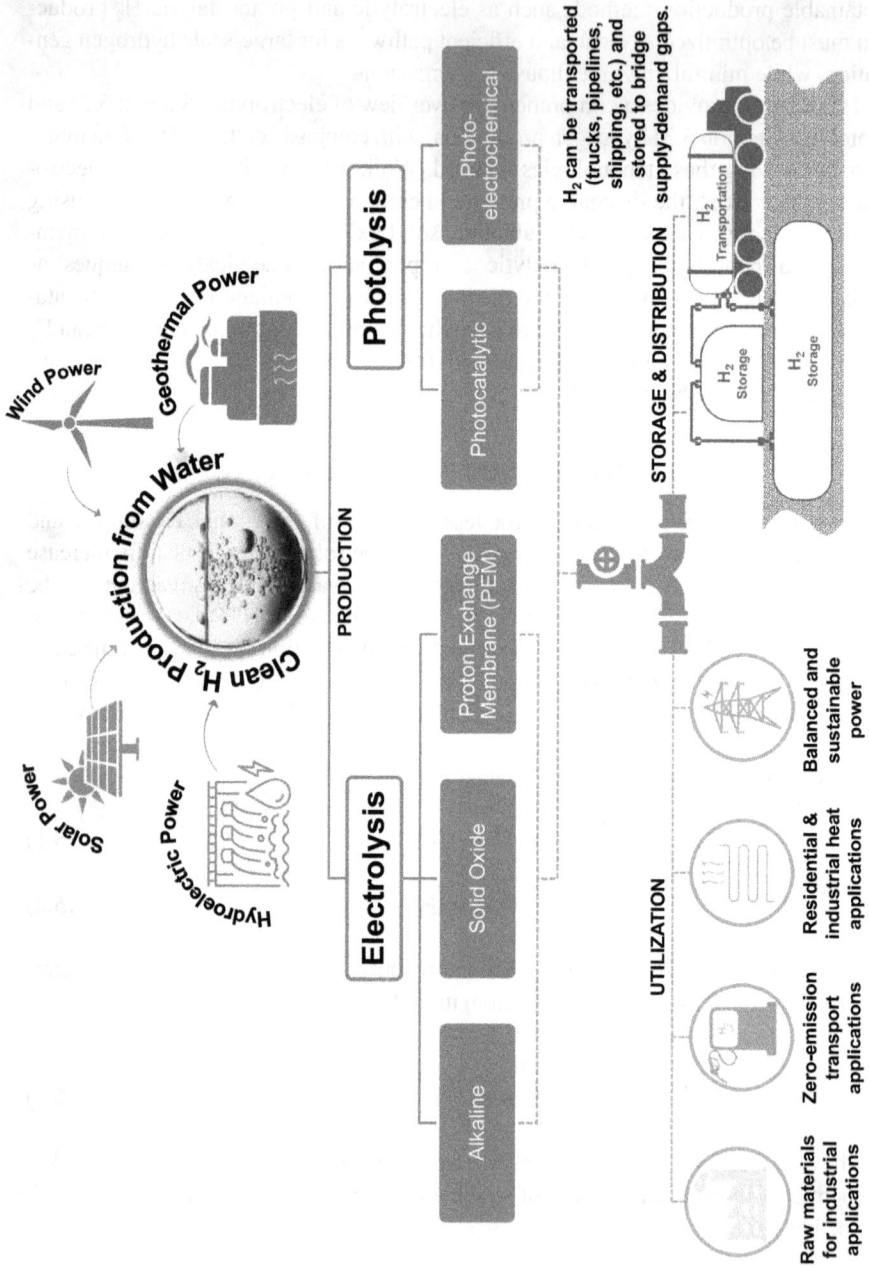

FIGURE 5.1 Graphical overview of clean hydrogen (H_2) flows from electrolytic and photolytic production methods to utilization.

For alkaline and proton exchange membrane (PEM) cells, these reactions occur in similar ways, albeit with slight variations. However, for solid oxide electrolyzers (SOE), Volmer is almost always followed by Tafel.

The rate-limiting step in the Tafel reaction is often the formation of the hydrogen molecule, and its kinetics are influenced by the coverage of the adsorbed hydrogen on

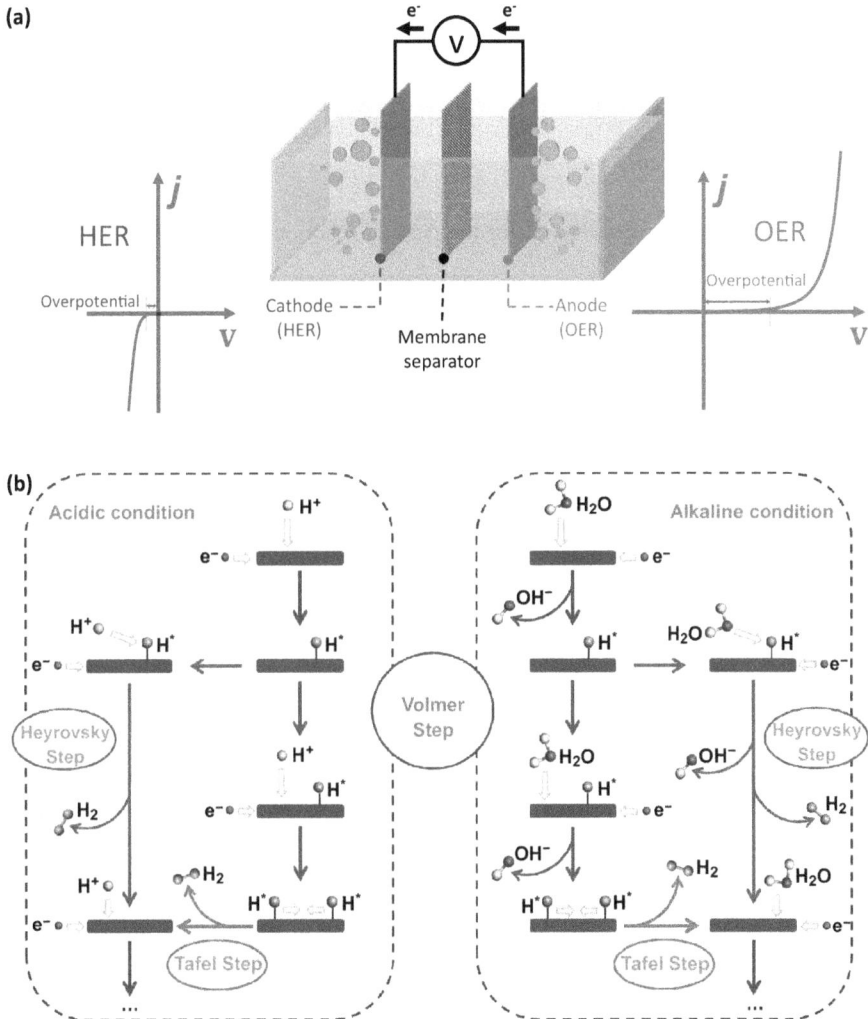

FIGURE 5.2 (a) A simplified graphical representation of hydrogen and oxygen evolution reactions that occur during electrolytic water splitting. (b) Schematic of the Volmer, Tafel, and Heyrovsky reaction pathways for hydrogen evolution reaction under acidic and alkaline conditions.

the electrocatalyst surface. The Heyrovsky reaction is favored at lower overpotentials and can be significant under alkaline conditions, resulting in a relatively faster reaction step compared to the Tafel reaction at low overpotentials. Ultimately, the occurrence of Heyrovsky or Tafel reactions after the Volmer reaction in the HER depends on factors such as the applied potential, electrocatalyst properties, pH, surface coverage, and reaction kinetics. Thus, the overall efficiency of HER will depend on an interplay between Tafel and Heyrovsky reactions, along with other factors relating to the electrocatalyst and operating conditions. Undoubtedly, the selection of an appropriate electrocatalyst can be pivotal to H_2 production, determining the reaction kinetics, selectivity, stability, and scalability.

To improve large-scale electrolysis efficiency, we need materials that are abundant, stable, and reactive. In this regard, the practical considerations for making electrocatalysts are as follows [5]:

1. Low inherent overpotential for the desired reaction (like producing hydrogen or oxygen).
2. Large active surface area to easily interact with reactants and remove products (gases, liquids, ions).
3. High electrical conductivity.
4. Good chemical stability and compatibility with the electrolyte.
5. High stability under electrochemical conditions (i.e., not corroding at high potentials).
6. Good structural integrity, especially at elevated temperatures.

Undoubtedly, the selection of an appropriate electrocatalyst is pivotal to H_2 production, determining the reaction kinetics, selectivity, stability, and scalability. Equally significant is the choice of electrode materials and design, which can profoundly influence the overall performance and efficiency of the electrolysis process. For a comprehensive comparative overview of electrolysis technologies, including operational parameters, reactions, and applications, see Table 5.1.

It is worth noting that although Sections 5.2.1.1–5.2.1.3 focus more broadly on electrocatalyst, there are some areas of overlap in relation to electrodes, which may be modified to address the issues with various system resistances that impede the overall HER efficiency during electrolysis. Lowering these resistances, increasing the ionic conductivity, and the charge-carrying capacity are other routes to attain considerable improvement [47]. Ultimately, the effectiveness of noble metal-based nano-electrocatalysts for HER is dependent on their size, shape, and facet [6].

5.2.1 NANOMATERIALS FOR ELECTROLYTIC HYDROGEN PRODUCTION

Extensive research efforts are ongoing to address the widely documented issues with high overpotentials, slow OER kinetics, O_2 contamination of H_2, and the undesirable generation of active oxygen species, which jeopardize the durability of electrolyzers. Nanomaterials play a crucial role in enhancing electrolytic hydrogen production by

TABLE 5.1

Comparative Overview of Electrolysis Technologies for Hydrogen Production: Operational Parameters, Reactions, and Applications

	Electrolytic Technologies		
	Alkaline	PEM	Solid Oxide
Operational Temperatures (°C)	<100	50–80	600–1000
pH	13–14	0–7	N/A
Charge Carriers	OH⁻	H⁺	O²⁻
Primary Reactions	HER OER	HER OER	OER
Reaction Mechanisms	Volmer Tafel Heyrovksky	Volmer Tafel Heyrovksky	Volmer Tafel
Commercial Applications	Water electrolyzers, often within industrial settings, for large-scale hydrogen production	Including but not limited to transportation, portable devices, and various commercial and industrial contexts	Mainly employed for industrial H_2 production and for fuel cell applications.
Technological Barriers	Limited efficiency, electrode degradation, electrolyte compatibility, achieving scalability for larger-scale production	Managing membrane durability, reducing catalyst cost, enhancing system efficiency, ensuring scalability for industrial applications	Dealing with high operating temperatures, developing stable materials for electrodes, improving ion transport, addressing challenges related to scaling up
TRL	★ ★ ★	★ ★ – ★ ★ ★	★ – ★ ★
Reactions	★	★ ★ – ★ ★ ★	★ ★ ★
Relative Efficiency	★	★ ★ – ★ ★ ★	★ ★ ★
Relative System Complexity	★	★ ★	★ ★ ★
Relative Cost	★	★ ★	★ ★ ★

Note: HER, OER, and TRL are hydrogen evolution reaction, oxygen evolution reaction, and technology readiness level, respectively. ★, ★★, and ★★★ indicate low, medium and high for the respective parameters.

contributing high surface area, tunable electronic structure, and increased active sites [6]. The overall efficiency of water-splitting processes, including electrolysis, is still relatively low. High overpotentials are often required to drive the reactions, leading to increased energy consumption and reduced energy efficiency. Interestingly, the in-practice overpotential can be considerably lower than the theoretical minimum for a given reaction due to internal and thermodynamic resistances. Of course, an appropriate electrocatalyst plays a significant role in lowering the concentration overpotential.

5.2.1.1   Noble Metal Nano-electrocatalysts

Noble metal catalysts like platinum (Pt) and Pt-group metals (PGMs) are widely reported to exhibit low overpotential; their cost and scarcity make full exploitation impractical. Thus, lower-cost noble metals like palladium and ruthenium (Ru) have been extensively employed [7]. Nanostructured palladium has emerged as an attractive alternative to platinum (Pt) for the hydrogen evolution reaction (HER), driven by its comparable atomic size to Pt and cost-effectiveness. Palladium's capacity for H_2 adsorption from both gas and electrolytes occurs through distinct α- and β-phases, which depend on H_2 concentration. The catalytic efficiency hinges on the morphologies of palladium nanoparticles (Pd NPs), shaped by thermodynamic or kinetic controls during crystal growth. Different shapes, such as octahedrons or cuboctahedrons can arise based on seed crystallinity and facet growth rates, resulting in diverse electronic and geometric structures. Octahedral Pd NPs, for instance, exhibit remarkable H_2 loading due to their small size and morphology. Additionally, 3D porous Pd NP assemblies hold potential for catalyzing hydrogen evolution and oxygen reduction reactions. Integration of components like $FeOx(OH)_{2-2x}$ in core-shell designs notably boosts HER activity, especially in alkaline conditions. Ruthenium also yields low H_2 desorption barrier, and like palladium, lends itself to morphological tailoring on the nanoscale. In fact, both materials can benefit greatly from the use of an appropriate support material to realize enhanced porosity, stability, surface area and conductivities (thermal and electrical). Ruthenium, in particular, has been shown to exhibit crystallinity-dependence in its catalytic behavior, which can be used for additional morphological control of performance. Enhanced stability may also be realized by nanocomposite synthesis based on stabilizing architectures such as core-shell structures [8]. Nonetheless, some inherent drawbacks must be considered for ruthenium, such as its higher attraction to OH than Pt, which usually slows down the HER rate, and its susceptibility to dissolution and aggregation-induced degradation. [3]. Notwithstanding, noble metal electrocatalysts demonstrate exceptional activity across a broad range of pH values for both the HER and OER [8].

5.2.1.2   Non-noble Metal Nano-electrocatalysts

Non-noble metal alternatives, such as nickel and cobalt, have garnered significant attention in this field. Extensive research has focused on first-row transition metal oxides, including Ni, Co, Mn, and Fe, owing to their favorable OER activity [9]. Investigating complex oxides with multiple constituent elements to fine-tune compositions for improved OER efficiency is highly significant. These metals offer a promising avenue for advancing sustainable hydrogen production. Furthermore, transition-metal carbides (TMCs), transition-metal sulfides (TMSs), transition-metal nitrides (TMNs), and transition-metal phosphides (TMPs) have emerged as noteworthy candidates to facilitate this goal [6, 9, 10]. Notably, some of these materials, including TMCs [11] and TMSs [12–14], exhibit remarkable bifunctional electrocatalytic performance, showcasing their potential to drive both HER and OER processes. However, it is important to acknowledge that each of these material classes comes with its own set of challenges. TMCs may experience corrosion susceptibility and necessitate complex synthesis procedures. In addition, TMSs, while promising, might face limitations in terms of electrical conductivity and could undergo sulfur evolution under specific conditions. Furthermore, TMNs could exhibit restricted selectivity and susceptibility to catalyst poisoning, while TMPs might grapple with surface oxidation and stability concerns. Despite these drawbacks, research in this area holds the promise of overcoming these challenges and unlocking the full potential of non-noble metal electrocatalysts for sustainable H_2 production.

In light of sluggish OER kinetics, researchers have extensively explored perovskite oxides [10–17] and spinel oxides [15, 18, 19], which possess ABO_3 and AB_2O_4 structures, respectively. These materials offer the potential for targeted performance tuning in the realm of electrocatalysts, yet they inherently suffer from poor electrical conductivity and low electrochemical surface area. However, these limitations are effectively mitigated through the integration of conductive carbonaceous materials, leveraging synergistic effects to enhance their conductivity and surface areas.

The strategy of employing engineered 3D hierarchical heterostructures with conductive supports, such as carbon and nickel substrates, has emerged as a promising avenue to achieve enhanced performance at low overpotentials for both HER and OER [20, 21]. Furthermore, the utilization of decorated doped nano-electrocatalysts has also gained prominence, offering a pathway to further optimize and fine-tune catalytic performance in these crucial electrochemical processes [22].

5.2.1.3   Metal-free Nano-electrocatalysts

Owing to their excellent electronic conductivity and good stability, carbon-based nanomaterials represent some of the most sought after candidates for metal-free electrocatalysts. Nonetheless, the performance of single-phased carbon materials is constrained by their inherent weaker material properties in terms of electrocatalytic ability. Despite the achievements attained by integrating precious metals and metal oxides into carbon catalysts for electrochemical applications, the significant cost and constrained availability of noble metals have emerged as pressing issues requiring resolution. Consequently, the manipulation of carbon supports through the introduction of heteroatoms (like N, O, F, Si, Se, B, P, and S) to modify their surface and physicochemical traits has gained attention in recent years. Heteroatoms can be used to confer electron donor properties, facilitating π-bonding – a pivotal factor that contributes to heightened stability, accelerated electron transfer rates, and enhanced durability [23–25]. Some excellent carbonaceous nanomaterials of interest are based on graphene quantum dots [26–28], graphene aerogels [29–33], graphene nanosheets [26, 34–36], and carbon nanotubes [37–40]. Notably, graphene-based nanocatalysts and supports exhibit good functionality in both alkaline and acid media. Other schemes for enhancing performance include defect and edge engineering and surface functionalization [19, 41].

An overview of some commonly employed materials from Sections 5.2.1.1 to 5.2.1.3 is presented in Table 5.2.

TABLE 5.2

Overview of Nanomaterials Employed in Electrocatalysts for Electrolytic Hydrogen Production

Nanomaterials	Electrolyte	Over-potential (mV)	Tafel Slope (mV/dec)	Current Density (mA/cm²)
Ru@C2N	0.5 M H₂SO₄	13	30	10
	1.0 M KOH	17	30	10
Co-Ru NSs	1 M KOH	13	29	10
N-Pt-Ni	1 M KOH	13	76	10
Pt₃.₂₁Ni@Ti₃C₂	0.5 M H₂SO₄	18	13	10
Pt@Ni@Co	0.1 M KOH	22	-	10
Mo₂TiC₂Tx–PtSA	0.5 M H₂SO₄	30	116	10
Pt/Ti₃C₂Tx-550	0.5 M H₂SO₄	32	32	10
Pd/Nb₂C–HF	0.5 M H₂SO₄	34	34	10
Ni₀.₉Co₀.₁@Nb-doped Ti₃C₂Tx	1.0 M KOH	43	116	10
CoS-P/CNT	0.5 M H₂SO₄	48	55	10
Pt-Ni3N	1 M KOH	50	36.5	10
Ru@CN-0.16	1 M KOH	50	53	10
Ti₃C₂Tx@Pt/SWCNTs	0.5 M H₂SO₄	62	78	10
Fe0.54Co0.46S0.92/CNTs/CC	1 M KOH	70	55	10
CoP/Ti₃C₂	0.5 M H₂SO₄	71	57	10
Ru/C₃N₄/C	0.1 M KOH	79	61	10
Cu-Mo-S	0.5 M H₂SO₄	96	52	10
MoS₂–Ti₃C₂	0.5 M H₂SO₄	98	45	10
N-S/C	0.5 M H₂SO₄	100	57	10
RuO2-NiO	6.0 M KOH	100	38.5	100
Ru NPs/NC	1 M KOH	112	31	10
Mo₂CTx/2H-MoS₂	0.5 M H₂SO₄	119	60	10
MoS2/rGO	0.5 M H₂SO₄	120	41	10
CuNDs/Ni3S2 NTs-CFs	1 M KOH	128	72.6	10
S-MoS2@C	0.5 M H₂SO₄	136	-	10
N–Ti₃C₂Tx-35	0.5 M H₂SO₄	162	69	10
P₃–V₂CTx	0.5 M H₂SO₄	163	74	10
Ti₃C₂ NFs	0.5 M H₂SO₄	169	97	10
MoS₂@Mo₂CTx	1.0 M KOH	176	207	10
P–Mo₂C/Ti₃C₂@NC	0.5 M H₂SO₄	177	57	10
N-WC	0.5 M H₂SO₄	190	149	200
MD-Ti₃C₂/MoSx	0.5 M H₂SO₄	196	41	50
N–Ti₃C₂Tx@600	0.5 M H₂SO₄	198	92	10
1T MoS2	0.5 M H₂SO₄	200	40	10
O-MoS2	0.5 M H₂SO₄	200	55	10
V₄C₃Tx	0.5 M H₂SO₄	200	168	10

(*Continued*)

TABLE 5.2   (Continued)
Overview of Nanomaterials Employed in Electrocatalysts for Electrolytic Hydrogen Production

Nanomaterials	Electrolyte	Over-potential (mV)	Tafel Slope (mV/dec)	Current Density (mA/cm^2)
NiFe-LDH/MXene/NF	1.0 M KOH	205	70	500
N-S/rGO	0.5 M H$_2$SO$_4$	270	81	10
Nb$_4$C$_3$T	1.0 M KOH	398	122	10
Ni$_3$FeN/rGO	1.0 M KOH	1530	-	10
NiFeP/3D strutted GN	1.0 M KOH	1540	-	10
NiQD@NC@rGO	1.0 M KOH	1563	-	10
NiSe/N-GN	1.0 M KOH	1600	-	10
Ni$_2$P/rGO	1.0 M KOH	1610	-	10
Ni$_2$Co$_2$P$_2$/GQDs	1.0 M KOH	1680	-	100
Ni-MoxC/N-GN	1.0 M KOH	1720	-	10

Data sourced from [7, 41–44].

5.2.2   CHALLENGES OF NANOMATERIAL-BASED ELECTROLYTIC PROCESSES FOR RENEWABLE H$_2$ PRODUCTION

Despite the exceptional potential for enhancing the operational efficiency and cost effectiveness of electrolytic H$_2$ production using nanomaterials, several challenges and open research problems remain to be addressed.

For nanomaterial-based electrocatalysts, some of the most pressing concerns that limit their practical applicability are around stability and long-term durability. While they typically exhibit high activity initially, the effects of exposure to electrochemical corrosion, sintering, and surface poisoning can render them susceptible to degradation over time [5, 45]. It is generally accepted that higher current densities can lead to increased degradation of the electrocatalyst over time due to the concomitant increases in temperature, gas evolution, and changes in pH [5].

Perhaps these incremental improvements in durability suggest the potential for further enhancements through the combination of favorable supporting matrices with hybrid nanocomposite constituents. Of course, among the many extensively employed routes for tailoring the electrocatalytic performance, some of the most promising have been doping [45–48], defect manipulation [49, 50], hybrid material synthesis [51–53], integration of single-atom catalytic sites, and the orchestrated interplay of these strategies [36, 48].

For HER, in alkaline and acidic media, durability runs in the order of thousands of cycles, and there is often a correlation between their degradation rate and the current density. Nonetheless, some relatively recent advancements have shown the possibilities for enhancing the durability of catalysts through careful design. Xue et al [45] reported favorably superior retention of catalytic activity for their CoNC/GD catalyst after 36,000, 38,000 and 9,000 cycles under basic, acidic, and neutral operating conditions, respectively. Similarly, Yang et al. [46] designed a FeCO-based nano-electrocatalyst to exceed 10k cycles with no critical losses in catalytic activity. In such a configuration, the use of carbon-based nanomaterials such as graphdiyne and metal-organic framework (MOF)-derived graphene layers in this pursuit, together with appropriate dopants such as nitrogen and carbon can likely yield highly synergistic effects. Additionally, atomic nanoconfinement appears to play a key protective role in mitigating corrosion-induced degradation, while dopants enabled an increase in the ionic adsorption sites for enhanced HER activity.

The use of heteroatoms to tailor the electrocatalytic performance of carbonaceous materials, as described in Section 5.2.1.3, remains a highly attractive scheme. Nevertheless, the immediate one-step creation of high-surface-area, metal-free carbon-based catalysts composed solely of C, H, O, and N necessitates further advancement. Innovations in synthetic methodologies and material design are key to realizing this ambitious goal.

While carbonaceous nanomaterials are swiftly garnering attention as electrocatalysts for water electrolysis, with a notable focus on alleviating sluggish OER (Oxygen Evolution Reaction) activity, further investigations are imperative to attain comprehensive fundamental insights into the underlying mechanisms governing OER activity, specifically concerning multi-walled CNT (carbon nanotube)-based systems, which despite exhibiting inferior electrical properties to their single-walled counterparts offer excellent thermal stability and can be synthesis relatively inexpensively [54–56]. The intricacies of the charge transfer process, facilitated by electron tunneling, demand thorough exploration and elucidation. Other areas that must be addressed include:

- **Enhancing Research Focus**: Exploring multi-material systems of metal oxides and functionalized carbon holds substantial promise for scientific investigation. Dedicated research endeavors to comprehend the intricate synergies between these materials will be both timely and valuable.
- **Optimizing Nanoparticle Size**: To enhance reactivity, it is imperative to engineer nanoparticles in the range of 5–10 nm. These dimensions facilitate an increased density of reactive surface sites. As such, the pursuit of novel synthesis techniques becomes imperative, enabling the production of smaller nanoparticles that markedly amplify the electroactive properties of these materials.
- **Electrocatalyst Poisoning**: Electrocatalysts can be deactivated or poisoned by the presence of certain species in the electrolyte. Understanding and mitigating these poisoning effects as newer nanoelectrocatalysts emerge is important for maintaining long-term catalytic activity.
- **Addressing Synthesis Challenges**: The creation of smaller nanoparticles necessitates innovative synthesis methodologies. To address this challenge, researchers must focus on developing new synthesis approaches capable of yielding the desired diminutive particle size. This advancement will significantly elevate the electroactivity of the resultant materials, ushering in transformative possibilities for their applications.

- **Scaling Up Electrochemical Processes**: Transitioning from laboratory-scale electrochemical reactions to large-scale industrial processes can pose significant challenges. Ensuring consistent electrode performance, managing heat and mass transfer, and addressing safety concerns are important considerations.

5.3 PHOTOLYTIC HYDROGEN PRODUCTION

Photolytic hydrogen production, that is, the use of photons (light energy) to drive the splitting of water into H_2 and O_2 gases involves the use of light-absorbing materials such as semiconductor photocatalysts or photoelectrodes to harness light energy and to initiate hydrogen and oxygen evolution reactions [57, 58]. Broadly speaking, photolysis in the context of water-splitting H_2 production can be assumed to encompass both photocatalytic and photoelectrochemical methods for which semiconductor photocatalysts and photoelectrodes are employed, respectively. For the former, H_2 is produced due to the catalytic activity on the reactants in the presence of light, whereas, in the case of the latter, the photoelectrode generates electron-hole pairs and drives HER and OER at the cathode and anode, respectively within an electrochemical cell. Some key differences are presented in Table 5.3. Nonetheless, this chapter will primarily focus on the use of nanomaterials in photocatalysis for H_2 production.

TABLE 5.3

An Overview of the Key Distinguishing Attributes of Photocatalytic and Photoelectrochemical H_2 Production Methods

	Photocatalytic H_2 Production	Photoelectrochemical H_2 Production
Energy Source	Light energy (photons)	Light energy (photons)
Role of Material	Separate photocatalyst material	Light-absorbing material integrated within an electrochemical cell
Redox Reactions	Occurs on the surface of the photocatalyst	Takes place at the photoelectrode within an electrochemical cell
H2 Generation	Catalyst facilitates H_2 evolution from water	Electrons from the photoelectrode drive H_2 evolution from water
O2 Generation	O_2 gas is a byproduct of water oxidation	O_2 gas is a byproduct of oxygen evolution reaction at the anode
Typical Catalyst	Semiconductor photocatalysts such as TiO_2 and MoS_2 are commonly used	Light-absorbing semiconductors such as Si and GaN are the most commonly used
Application	Suitable for distributed or decentralized setups	Suitable for centralized or large-scale systems
Technology Readiness Level (TRL)	Lower TRL, mainly in research and laboratory scale	Higher TRL, with some pilot and demonstration projects

5.3.1 NANOMATERIALS FOR PHOTOCATALYTIC HYDROGEN PRODUCTION

In the quest for sustainable and efficient photoelectrochemical hydrogen production, the search for ideal semiconductor photocatalysts has become paramount. These photocatalysts are responsible for absorbing light and initiating the water splitting process. To harness the full potential of these materials, several key characteristics must be carefully considered [59]:

1. **Optimal Bandgap Energy**: The ideal semiconductor photocatalyst should possess an optimal bandgap (BG) energy that aligns with the solar spectrum. This alignment ensures maximum light absorption, covering both visible and near-infrared regions effectively.
2. **Strategic Energy Level Positioning**: The conduction band (CB) and valence band (VB) energy levels should be strategically positioned to enable effective charge separation.
   a. The CB energy level must be more negative than the redox potential of the hydrogen evolution reaction (HER).
   b. The VB energy level should be more positive than the redox potential of the oxygen evolution reaction (OER).

This arrangement guarantees that photogenerated electrons are readily available for H_2 reduction, while holes can be efficiently utilized for oxygen evolution. The principle for photocatalytic H_2 production via water splitting is shown in Figure 5.3.

Beyond bandgap and band positions, charge carrier mobility is a critical factor in semiconductor photocatalysts. High mobility ensures minimal recombination of photogenerated electron-hole pairs, thereby enhancing the catalyst's overall efficiency. Furthermore, achieving a high surface area through nanostructuring offers numerous benefits, including increased light absorption, improved charge separation, and more active sites for catalytic reactions. Stability and durability are equally

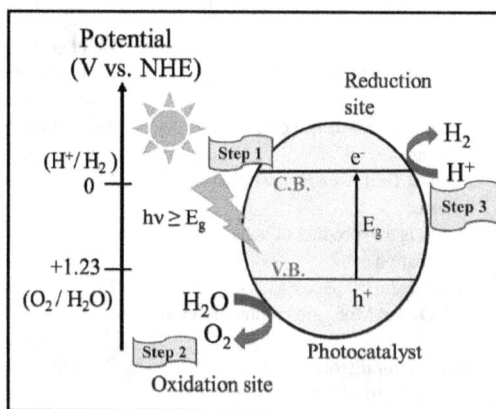

FIGURE 5.3 The stepwise mechanism for photocatalytic water splitting.

(Reproduced from [56] with permission from Elsevier.)

paramount, as the photocatalyst must withstand prolonged exposure to light, water, and reactive intermediates without compromising performance. Moreover, the semiconductor photocatalyst should exhibit robust catalytic activity towards both hydrogen evolution reaction (HER) and oxygen evolution reaction (OER) to drive the water-splitting process efficiently.

Nanomaterials have emerged as indispensable components in the realm of photocatalytic hydrogen production. Their main advantage lies in their amplified surface area, which provides ample active sites to foster photochemical reactions. Exploiting the high surface area-to-volume ratio of nanomaterials can facilitate efficient charge separation and transfer, mitigating the undesired recombination of photoexcited charge carriers. Furthermore, their ability to harness a broader range of the solar spectrum, including UV and visible light, is key to driving the water-splitting reaction and unlocking a sustainable pathway for hydrogen generation.

The tunability of nanomaterials' bandgap constitutes a decisive asset in photocatalytic systems. Engineering nanophotocatalyst enables the precise matching of the bandgap with the photon energies in the solar spectrum, which maximizes light absorption and the subsequent generation of photo-excited charge carriers. The chemical stability and durability exhibited by some nanomaterials guarantee sustained photocatalytic activity, ensuring prolonged performance under demanding conditions. As effective catalyst supports, nanomaterials offer enhanced dispersion and interaction with catalytic centers, augmenting their effectiveness in promoting H_2 evolution.

Carbides [60], sulfides [61–63], and oxides [64–66] are some of the most commonly employed photocatalysts for H_2 production today, and as research continues to advance in this area, metallic and non-metallic nanomaterials are being actively explored. With a plethora of nanomaterials in the field, nanophotocatalysts can be designed with the particular reaction parameters and conditions in mind.

Metal oxides like titanium dioxide (TiO_2) nanoparticles offer remarkable electron-hole separation efficiency and wide bandgap, which makes them ideal candidates for UV-driven water-splitting H_2 production methods. Similarly, zinc oxide (ZnO) nanorods with their elongated structures, are excellent for photocatalysis by visible light, offering abundance active sites, while enabling exceptional bandgap tenability and charge transport. Other metal oxides include SnO_2 and CeO_2. The success of the hydrogen production process using these metal oxide-based catalysts relies on various factors, such as the crystallinity phase and size of the materials, the effective surface area available for catalysis, the pore size and structure, as well as the adsorption capacity of the catalysts.

Notable non-metallic oxides like graphitic carbon nitride (g-C_3N_4) nanosheets also exhibit bandgap tenability and efficient charge separation.

There are, of course, intriguing 3D nanomaterials in the field like metal-organic frameworks (MOFs), which consist of intricate, 3D, porous nano-architectures with tunable metal centers. These MOFs have been extensively researched for UV-, visible light-, and NIR (near-infrared)-assisted H_2 production [7, 67].

Alongside their potential application as photocatalysts, carbon-based nanomaterials have demonstrated effectiveness as cocatalysts and support materials. This application enhances catalytic activity by augmenting adsorption and active sites, leading to a reduced bandgap and intensifying photosensitization. Additionally, these nano-materials can act as electron acceptors and facilitate the transportation of electrons during photocatalytic processes [56].

Several nanomaterials are currently employed for photocatalytic H_2 production, including SiC, TiO_2, CdS, and $CuInSe_2$. These materials can be broadly categorized as metal-oxide (e.g., TiO_2 and WO_2) and non-metal oxide (e.g., g-C_3N_4, B_4C, $CuInSe_2$, and CdS) photocatalysts [68]. However, the present achievable efficiencies in H_2 production (solar-to-fuel) using these photocatalysts remain low, and are still far from meeting the operational requirements for industrial applications [69].

Advancements in research within the realm of photocatalytic water splitting have resulted in the creation of over 100 new catalysts, encompassing carbides, sulfides, and oxides [70]. Despite these strides, various challenges persist that hinder their practical industrial use. These challenges encompass issues like the limited reactivity of photocatalysts to ultraviolet light (resulting in low visible-light quantum effi-ciency), the rapid recombination of photogenerated charges, subpar durability and physicochemical stability, elevated toxicity, and difficulties in large-scale catalyst preparation [71–73]

The primary hurdles in photocatalysis revolve around inadequate light absorption and insufficient charge separation. These challenges stem primarily from the sub-stantial bandgaps inherent in photocatalysts, especially metal oxides. These wide bandgaps position these materials beyond the reach of visible light in the electro-magnetic spectrum, impeding their effective utilization of solar energy. In contrast, photocatalysts capable of harnessing visible light exhibit greater efficiency, given the significant role that visible light plays within the electromagnetic spectrum [68, 74, 75].

For instance, TiO_2, boasting a bandgap of 3.2 eV, achieves only 1% efficiency in converting solar energy to hydrogen (solar-to-H_2 conversion efficiency). Conversely, Fe_2O_3, with its bandgap of 2.2 eV, possesses the potential for a theoretical effi-ciency of 15% [68, 76]. This limitation stems from their inability to fully capture the energy present in the solar spectrum, despite their commendable catalytic properties. Nonetheless, photocatalysts proficient in utilizing visible light demonstrate enhanced efficiency and efficacy, which is in part, attributed to the prominent role of visible light within the electromagnetic radiation spectrum.

A significant challenge in the field revolves around the issues of inadequate light absorption and charge separation, primarily stemming from the substantial band-gaps present within photocatalysts [74, 75]. These considerable bandgaps, particu-larly prevalent in metal oxides, frequently position themselves outside the range of visible light within the electromagnetic spectrum [68]. Consequently, these materi-als struggle to tap into the solar spectrum's energy potential, regardless of their attractive catalytic attributes. In contrast, photocatalysts capable of effectively uti-lizing visible light demonstrate heightened efficiency and effectiveness, owing to the influential role played by visible light within the electromagnetic radiation spec-trum [74].

Some notable photocatalysts, include graphitic carbon nitrides, conjugated micro-porous polymers (CMPs), linear conjugated polymers (LCPs), and covalent organic frameworks [72, 74, 77]. Noteworthy among these are graphitic nitrides (g-C_3N_4), a subset of organic semiconductors, heralded as promising contenders for solar-driven H_2 production via water splitting [78]. Their optoelectronic attributes and reduced environmental impact in comparison to inorganic and metal-complex materials makes them highly desirable. However, the pronounced molecular bulk of g-C_3N_4 often leads to the formation of unwieldy particles in aqueous solutions, transcending the light's penetration length scale [72]. This inevitably gives rise to optical losses, at the detriment of catalytic efficacy. Hence, the particle size of the photocatalyst emerges as a pivotal determinant of performance.

Refining particle dimensions to the nanometric scale has been shown to improve photocatalytic performance [79]. Tailoring strategies, including nanostructure engineering, provide the means to derive thin-layer nanosheets from bulk g-C_3N_4 [71, 74]. Techniques span physical exfoliation, both liquid and chemical, emulsion polymerization, sonication, template synthesis, and the nanoprecipitation of polymer dots. Similarly, anisotropic 1D nanostructures are fast gaining research attention, with enhanced charge transport along their lengths. This directional transport can enable the migration of charge carriers to the surface of the nanostructure for improved efficiency. Noteworthy is the work of Zhu et al., who crafted a hierarchical 1D configuration involving Ni_3S_2 nanosheets supported on a carbon nanotube scaffold, resulting in heightened photocatalytic activity in water splitting for H_2 production [80]. Further afield, Chava et al. introduced 1D CdS–Au/MoS_2 hierarchical core/shell heteronano structures (CSHNSs) through a facile two-step hydrothermal route, revealing exceptional stability and efficiency in hydrogen production [81].

2D nanostructures have also attracted research interest due to their augmented specific area, increasing the number of reaction sites. Unlike their bulk photocatalyst counterparts, their shorter diffusion lengths limit the prevalence of charge carrier recombination [71]. Ganguly et al. offer excellent insights on 2D nanomaterials for photocatalytic H_2 production [73]. Of particular note, Zhang et al. synthesized a 2D nanophotocatalyst by integrating ultrathin TiO2 nanosheets doped with quantum Cu(II) nanodots (QCNs), establishing that tempering the recombination rate of photogenerated electron-hole pairs enhances the material's photocatalytic performance [82]. Meanwhile, Xu et al. reported a thirteen-fold amplification in the photocatalytic activity of g-C_3N_4 nanosheets post electrostatic assembly with hematite α-Fe_2O_3 nanoplates, wherein Pt served as a cocatalyst and triethanolamine played the role of a hole scavenger, culminating in performance enhancement [83]. A noteworthy proposition from Mahzoon et al. outlines a cost-effective and straightforward approach to fabricating a hybrid 1D-2D g-C_3N_4 heterojunction nanophotocatalyst [71]. In the present landscape, cocatalyst modifications is a promising route for improving performance [84].

Factors influencing photocatalyst conversion efficiency encompass various aspects such as the extent of light absorption, bandgap, exciton generation, and charge mobility. An efficient photocatalyst should possess a specific photoactivity exceeding 104 μmol/h/g and the capability to facilitate H–O bond cleavage by reducing the reaction energy barrier substantially. Moreover, the separation and recombination of photogenerated charge carriers determine the overall efficiency, along with the charge collection rate at electrodes. The structural attributes of the photocatalyst particles, including size, shape, morphology, and crystallinity, can also significantly impact their performance. In practical applications and under specific operating conditions, additional factors come into play. The photocatalyst's resistance and stability under visible light are essential, ensuring photohydrostability without undergoing photocorrosion in aqueous solutions. The nature of the cocatalyst and its preparation method, as well as the concentration of sacrificial agents, influence the overall reaction. Moreover, factors like pH and operating temperature, suitability for large-scale preparation, cost-effectiveness, and toxicity level need careful consideration. The size and shape tunability of photocatalyst particles allow customization for optimal performance. Furthermore, the removal of nanoarchitectures by photons, facilitating their arrival at the reactive sites, is crucial for efficient operation. These multifaceted factors collectively determine the success of a photocatalytic process, bridging the gap between theoretical potential and practical applicability [68, 73].

An overview of various comparative photocatalytic performances from the literature on nanomaterial usage is presented in Table 5.4.

5.3.2 Challenges of Nanomaterial-based Photocatalytic Processes for Renewable H_2 Production

While metal oxides are some of the most widely used photocatalysts in the field, owing to large bandgaps that lie outside the electromagnetic spectrum, their use is associated with unsuitable poor light absorption, insufficient charge separation, and an inability to utilize the solar spectrum [11].

TABLE 5.4
Comparison of Reported Nanomaterial-Based Photocatalytic Performance

Nanomaterials	Light Source	Sacrificial Agents	H_2 Evolution (Rounded Values)	Refs.
$CNTs/ZnFe_2O_4$	Mercury lamp	Methanol	18 μmol/h/g	[85]
TiO_2/functionalized CNTs	Solar light	Glycerol	2134 μmol/h/g	[86]
QD Bi_2O_3 TiO_2	Solar light	Glycerol/Water	26 μmol/h/g	[87]
QD TiO_2	Solar light	Glycerol/Water	4 μmol/h/g	[87]
QD $Ni(OH)_2/TiO_2$	Solar light	Crude glycerol/Water	4.71 μmol/h/g	[88]
QD $Ni(OH)_2/TiO_2$	Solar light	Crude glycerol/Water	45.57 μmol/h/g	[88]
$g\text{-}C_3N_4$	Visible light	Triethanolamine/Water	93–2134 μmol/h/g	[71]
$g\text{-}C_3N_4/WO_3$	Xenon lamp, visible light	Triethanolamine/Water	1853 μmol/h/g	[89]
TiO_2/MoS_2-Graphene	UV-visible	Ethanol Water	165.3 μmol/h/g	[63]

The pursuit of efficient photocatalytic water splitting for hydrogen generation presents a promising avenue to address the escalating energy needs. However, realizing this potential comes with several obstacles, necessitating solutions to enhance solar conversion efficiency. Challenges include optimizing spectral sensitivity to harness a broader range of solar light, mitigating charge recombination rates, and bridging the gap in overpotential values between semiconductor photocatalysts and water. Exploring the following avenues is imperative:

- **Harmonizing Cocatalyst Integration and Electron Transfer**: The seamless integration of cocatalysts onto nanomaterial surfaces stands as a vital catalyst for steering the surface reactions integral to hydrogen evolution. Attaining optimal cocatalyst loading, precise placement, and streamlined electron transfer pathways to and from nanomaterials bear the potential to greatly influence overall catalytic activity. The pursuit of methodologies for controlled cocatalyst deposition and electron transport remains an open problem.
- **Catalyst Stability and Durability**: Nanomaterials may undergo structural changes or degradation over extended photocatalytic reactions, leading to reduced stability and performance. The harsh conditions of water splitting, especially in the presence of reactive intermediates, can accelerate material degradation. Developing strategies to enhance the stability and durability of nanomaterials under photocatalytic conditions is a critical aspect that requires in-depth material understanding and innovative protective coatings.
- **Navigating Quantum Size Effects and Catalyst Poisoning**: Quantum effects can engender size-dependent characteristics in nanomaterials, exerting influence over electronic structure and catalytic activity. Additionally, certain nanomaterials could be vulnerable to poisoning by adsorbed species, exerting long-term impact on their performance. Mastery over these effects, coupled with effective countermeasures, holds the key to realizing consistent and dependable hydrogen production.
- **Bandgap Engineering and Light Harvesting**: Tailoring the bandgap of nanomaterials to match the solar spectrum is a fundamental challenge in photocatalysis. While narrow bandgap materials can absorb visible light effectively, they may also introduce challenges related to charge carrier dynamics, including recombination. Designing heterostructured nanomaterials and exploring alloying or doping strategies are avenues for achieving efficient light harvesting and charge separation.
- **Enhancing Light Absorption**: Novel strategies should be devised to amplify solar light absorption using economically viable and stable semiconductors with heightened achromatic response. The current semiconductors exhibit insufficient light absorption and suffer from electron-hole pair recombination. Various studies have tackled these hurdles by incorporating cations or anions through doping, thereby altering the bandgap of photocatalysts. This modification empowers them to endure the redox potential of water and advance the water-splitting process.

Ultimately, gaps in knowledge for the peculiar water-splitting mechanisms that emerge should be addressed, specifically concerning facets such as light absorption, charge segregation, charge mobility across semiconductor interfaces, and the fundamental stages underlying hydrogen evolution. By elucidating the fundamentals across various water-splitting methodologies, along with assessing the merits and demerits of diverse modification tactics, proficient photocatalytic configurations can be devised. This will culminate in harnessing an expansive solar spectrum at minimal expense, thus establishing solar water splitting as a viable and practical solution for meeting current and future energy demands.

5.4 CONCLUSIONS

In this chapter, the significance of nanomaterials in the context of electrolytic and photolytic hydrogen production has been highlighted. These advanced materials hold the promise of offering solutions to address prevailing limitations. When appropriately integrated for electrolytic production methods, improved efficiency and reduced overpotentials may be readily realised. Similarly, for photolytic routes, they can enable efficient utilization of solar energy for H_2 generation by harnessing the photoexcitation of semiconductor nanomaterials, initiating charge separation, and driving the catalytic splitting of water molecules into hydrogen and oxygen species. Addressing the nanomaterial-specific challenges identified within this work may unlock the full potential of nanomaterials and pave the way for efficient and sustainable photocatalytic hydrogen production, contributing to the advancement of clean energy technologies.

REFERENCES

[1] Goldman Sachs, "The hydrogen revolution accelerates," *Intelligence*, Mar. 17, 2023. https://www.goldmansachs.com/intelligence/pages/the-hydrogen-revolution-accelerates.html (accessed Jul. 17, 2023).

[2] International Energy Agency, "Global Hydrogen Review 2021," 2022. doi: 10.1787/39351842-en

[3] J. Zhu, L. Hu, P. Zhao, L. Y. S. Lee, and K. Y. Wong, "Recent advances in electrocatalytic hydrogen evolution using nanoparticles," *Chemical Reviews*, vol. 120, no. 2. 2020. doi: 10.1021/acs.chemrev.9b00248

[4] J. Wei et al., "Heterostructured electrocatalysts for hydrogen evolution reaction under alkaline conditions," *Nano-Micro Letters*, vol. 10, no. 4. 2018. doi: 10.1007/s40820-018-0229-x

[5] F. M. Sapountzi, J. M. Gracia, C. J. (Kees-Jan) Weststrate, H. O. A. Fredriksson, and J. W. (Hans) Niemantsverdriet, "Electrocatalysts for the generation of hydrogen, oxygen and synthesis gas," *Progress in Energy and Combustion Science*, vol. 58. 2017. doi: 10.1016/j.pecs.2016.09.001

[6] Mahmoud, A.E.D., et al., Facile synthesis of reduced graphene oxide by Tecoma stans extracts for efficient removal of Ni (II) from water: batch experiments and response surface methodology. *Sustainable Environment Research*, 2022. 32(1): p. 22.

[7] E. I. Epelle et al., "A comprehensive review of hydrogen production and storage: A focus on the role of nanomaterials," *International Journal of Hydrogen Energy*, vol. 47, no. 47, pp. 20398–20431, 2022, doi: 10.1016/J.IJHYDENE.2022.04.227

[8] J. Yu, Q. He, G. Yang, W. Zhou, Z. Shao, and M. Ni, "Recent advances and prospective in Ruthenium-based materials for electrochemical water splitting," *ACS Catalysis*, vol. 9, no. 11, 2019, doi: 10.1021/acscatal.9b02457

[9] Y. Cheng and S. P. Jiang, "Advances in electrocatalysts for oxygen evolution reaction of water electrolysis-from metal oxides to carbon nanotubes," *Progress in Natural Science: Materials International*, vol. 25, no. 6. 2015. doi: 10.1016/j.pnsc.2015.11.008

[10] S. Kumar et al., "Lanthanide based double perovskites: Bifunctional catalysts for oxygen evolution/reduction reactions," *International Journal of Hydrogen Energy*, vol. 46, no. 33, 2021, doi: 10.1016/j.ijhydene.2021.02.141

[11] J. Il Jung, H. Y. Jeong, J. S. Lee, M. G. Kim, and J. Cho, "A bifunctional perovskite catalyst for oxygen reduction and evolution," *Angewandte Chemie, International Edition*, vol. 53, no. 18, 2014, doi: 10.1002/anie.201311223

[12] M. Risch et al., "Structural changes of cobalt-based perovskites upon water oxidation investigated by EXAFS," *Journal of Physical Chemistry C*, vol. 117, no. 17, 2013, doi: 10.1021/jp3126768

[13] S. Raabe et al., "In situ electrochemical electron microscopy study of oxygen evolution activity of doped manganite perovskites," *Advanced Functional Materials*, vol. 22, no. 16, 2012, doi: 10.1002/adfm.201103173

[14] S. Gupta, W. Kellogg, H. Xu, X. Liu, J. Cho, and G. Wu, "Bifunctional perovskite oxide catalysts for oxygen reduction and evolution in alkaline media," *Chemistry – An Asian Journal*, vol. 11, no. 1, 2016, doi: 10.1002/asia.201500640

[15] J. X. Flores-Lasluisa, F. Huerta, D. Cazorla-Amorós, and E. Morallón, "Transition metal oxides with perovskite and spinel structures for electrochemical energy production applications," *Environmental Research*, vol. 214, 2022, doi: 10.1016/j.envres.2022.113731

[16] K. J. May et al., "Influence of oxygen evolution during water oxidation on the surface of perovskite oxide catalysts," *Journal of Physical Chemistry Letters*, vol. 3, no. 22, 2012, doi: 10.1021/jz301414z

[17] J. Sun et al., "A bifunctional perovskite oxide catalyst: The triggered oxygen reduction/evolution electrocatalysis by moderated Mn-Ni co-doping," *Journal of Energy Chemistry*, vol. 54, 2021, doi: 10.1016/j.jechem.2020.05.064

[18] M. Cai et al., "Three-dimensional and in situ-activated spinel oxide nanoporous clusters derived from stainless steel for efficient and durable water oxidation," *ACS Applied Materials & Interfaces*, vol. 12, no. 12, 2020, doi: 10.1021/acsami.0c00701

[19] G. A. Gebreslase, D. Sebastián, M. V. Martínez-Huerta, and M. J. Lázaro, "Nitrogen-doped carbon decorated-Ni3Fe@Fe3O4 electrocatalyst with enhanced oxygen evolution reaction performance," *Journal of Electroanalytical Chemistry*, vol. 925, 2022, doi: 10.1016/j.jelechem.2022.116887

[20] H. Zhang, A. W. Maijenburg, X. Li, S. L. Schweizer, and R. B. Wehrspohn, "Bifunctional heterostructured transition metal phosphides for efficient electrochemical water splitting," *Advanced Functional Materials*, vol. 30, no. 34. 2020. doi: 10.1002/adfm.202003261

[21] S. Wang, A. Lu, and C. J. Zhong, "Hydrogen production from water electrolysis: role of catalysts," *Nano Convergence*, vol. 8, no. 1. 2021. doi: 10.1186/s40580-021-00254-x

[22] M. Fu et al., "Fe_2O_3 and Co bimetallic decorated nitrogen doped graphene nanomaterial for effective electrochemical water split hydrogen evolution reaction," *Journal of Electroanalytical Chemistry*, vol. 849, 2019, doi: 10.1016/j.jelechem.2019.113345

[23] M. A. Wahab et al., "Nanoconfined synthesis of nitrogen-rich metal-free mesoporous carbon nitride electrocatalyst for the oxygen evolution reaction," *ACS Applied Energy Materials*, vol. 3, no. 2, 2020, doi: 10.1021/acsaem.9b01876

[24] Y. Meng, X. Huang, H. Lin, P. Zhang, Q. Gao, and W. Li, "Carbon-based nanomaterials as sustainable noble-metal-free electrocatalysts," *Frontiers in Chemistry*, vol. 7. 2019. doi: 10.3389/fchem.2019.00759

[25] Y. Zhang et al., "Fabrication of 2D ordered mesoporous carbon nitride and its use as electrochemical sensing platform for H2O2, nitrobenzene, and NADH detection," *Biosensors & Bioelectronics*, vol. 53, 2014, doi: 10.1016/j.bios.2013.10.001

[26] A.E.D. Mahmoud, Graphene-based nanomaterials for the removal of organic pollutants: Insights into linear versus nonlinear mathematical models. *Journal of Environmental Management*, 2020. 270: p. 110911.

[27] R. Liu et al., "N-doped vertical graphene arrays/carbon quantum dots derived from vinegar residue as efficient water-splitting catalyst in a wide pH range," *Colloids and Surfaces A: Physicochemical and Engineering Aspects*, vol. 655, 2022, doi: 10.1016/j.colsurfa.2022.130258

[28] A.E.D. Mahmoud, et al., "Mechanochemical versus chemical routes for graphitic pre-cursors and their performance in micropollutants removal in water." *Powder Technology*, 2020. 366: p. 629–640.

[29] Y. Shi, W. Dai, M. Wang, Y. Xing, X. Xia, and W. Chen, "Bioinspired construction of Ruthenium-decorated nitrogen-doped graphene aerogel as an efficient electrocatalyst for hydrogen evolution reaction," *Chemical Research in Chinese Universities*, vol. 36, no. 4, 2020, doi: 10.1007/s40242-020-0167-2

[30] S. H. Koo et al., "Cobalt based nanoparticles embedded reduced graphene oxide aerogel for hydrogen evolution electrocatalyst," *Particle and Particle Systems Characterization*, vol. 36, no. 7, 2019, doi: 10.1002/ppsc.201900090

[31] X. Fu et al., "Co-N decorated hierarchically porous graphene aerogel for efficient oxygen reduction reaction in acid," *ACS Applied Materials & Interfaces*, vol. 8, no. 10, 2016, doi: 10.1021/acsami.5b12746

[32] Y. Li et al., "Tungsten oxide/reduced graphene oxide aerogel with low-content platinum as high-performance electrocatalyst for hydrogen evolution reaction," *Small*, vol. 17, no. 37, 2021, doi: 10.1002/smll.202102159

[33] N. K. A. Venugopal et al., "Prussian blue-derived iron phosphide nanoparticles in a porous graphene aerogel as efficient electrocatalyst for hydrogen evolution reaction," *Chemistry - An Asian Journal*, vol. 13, no. 6, 2018, doi: 10.1002/asia.201701616

[34] V. Thirumal et al., "Nitrogen and nitrogen-sulfur doped graphene nanosheets for efficient hydrogen productions for HER studies," *International Journal of Hydrogen Energy*, vol. 47, no. 98, 2022, doi: 10.1016/j.ijhydene.2022.06.136

[35] Mahmoud, A.E.D., A. Stolle, and M. Stelter, Sustainable synthesis of high-surface-area graphite oxide via dry ball milling. ACS Sustainable Chemistry & Engineering, 2018. 6(5): p. 6358–6369.

[36] M. B. Z. Hegazy, M. R. Berber, Y. Yamauchi, A. Pakdel, R. Cao, and U. P. Apfel, "Synergistic electrocatalytic hydrogen evolution in Ni/NiS nanoparticles wrapped in multi-heteroatom-doped reduced graphene oxide nanosheets," *ACS Applied Materials & Interfaces*, vol. 13, no. 29, pp. 34043–34052, Jul. 2021, doi: 10.1021/acsami.1c05888

[37] Y. Zhang et al., "Pyridine-grafted nitrogen-doped carbon nanotubes achieving efficient electroreduction of CO2 to CO within a wide electrochemical window," *Journal of Materials Chemistry A*, vol. 10, no. 4, pp. 1852–1860, Jan. 2022, doi: 10.1039/d1ta09491b

[38] W. Zhao et al., "Temperature differentiated synthesis of hierarchically structured N,S-Doped carbon nanotubes/graphene hybrids as efficient electrocatalyst for hydrogen evolution reaction," *Journal of Alloys and Compounds*, vol. 848, 2020, doi: 10.1016/j.jallcom.2020.156528

[39] D. Xia, X. Yang, F. Kang, J. Li, and L. Gan, "Development of Fe/N-doped carbon nanotubes as a stable non-precious electrocatalyst for oxygen reduction reaction," *ECS Meeting Abstracts*, vol. MA2019-02, no. 35, 2019, doi: 10.1149/ma2019-02/35/1633

[40] A. Ali, Y. Liu, R. Mo, P. Chen, and P. K. Shen, "Facile one-step in-situ encapsulation of non-noble metal Co2P nanoparticles embedded into B, N, P tri-doped carbon nanotubes for efficient hydrogen evolution reaction," *International Journal of Hydrogen Energy*, vol. 45, no. 46, 2020, doi: 10.1016/j.ijhydene.2020.06.235

[41] A. Ali and P. K. Shen, "Recent progress in graphene-based nanostructured electrocatalysts for overall water splitting," *Electrochemical Energy Reviews*, vol. 3, no. 2. 2020. doi: 10.1007/s41918-020-00066-3

[42] Z. Kang et al., "Recent progress of MXenes and MXene-based nanomaterials for the electrocatalytic hydrogen evolution reaction," *Journal of Materials Chemistry A*, vol. 9, no. 10. 2021. doi: 10.1039/d0ta11735h

[43] Y. Zheng et al., "High electrocatalytic hydrogen evolution activity of an anomalous ruthenium catalyst," *Journal of the American Chemical Society*, vol. 138, no. 49, 2016, doi: 10.1021/jacs.6b11291

[44] J. Mahmood et al., "An efficient and pH-universal ruthenium-based catalyst for the hydrogen evolution reaction," *Nature Nanotechnology*, vol. 12, no. 5, 2017, doi: 10.1038/nnano.2016.304

[45] Y. Xue et al., "Extraordinarily durable graphdiyne-supported electrocatalyst with high activity for hydrogen production at all values of pH," *ACS Applied Materials & Interfaces*, vol. 8, no. 45, 2016, doi: 10.1021/acsami.6b12655

[46] Y. Yang, Z. Lun, G. Xia, F. Zheng, M. He, and Q. Chen, "Non-precious alloy encapsulated in nitrogen-doped graphene layers derived from MOFs as an active and durable hydrogen evolution reaction catalyst," *Energy & Environmental Science*, vol. 8, no. 12, 2015, doi: 10.1039/c5ee02460a

[47] Z. Ren et al., "Tungsten-doped CoP nanoneedle arrays grown on carbon cloth as efficient bifunctional electrocatalysts for overall water splitting," *ChemElectroChem*, vol. 6, no. 20, 2019, doi: 10.1002/celc.201901417

[48] Y. Zheng et al., "Toward design of synergistically active carbon-based catalysts for electrocatalytic hydrogen evolution," *ACS Nano*, vol. 8, no. 5, 2014, doi: 10.1021/nn501434a

[49] Z. H. Tan, X. Y. Kong, B. J. Ng, H. Sen Soo, A. R. Mohamed, and S. P. Chai, "Recent advances in defect-engineered transition metal dichalcogenides for enhanced electrocatalytic hydrogen evolution: Perfecting imperfections," *ACS Omega*, 2022, doi: 10.1021/acsomega.2c06524

[50] D. K. Perivoliotis, J. Ekspong, X. Zhao, G. Hu, T. Wågberg, and E. Gracia-Espino, "Recent progress on defect-rich electrocatalysts for hydrogen and oxygen evolution reactions," *Nano Today* 2023, doi: 10.1016/j.nantod.2023.101883

[51] J. Hou, Y. Sun, Y. Wu, S. Cao, and L. Sun, "Promoting active sites in core-shell nanowire array as Mott-Schottky Electrocatalysts for efficient and stable overall water splitting," *Advanced Functional Materials*, vol. 32, no. 33, 2022, doi: 10.1002/adfm.202206634

[52] Anandhi, P., et al., "The enhanced energy density of rGO/TiO2 based nanocomposite as electrode material for supercapacitor." *Electronics* 2022. 11(11): p. 1792.

[53] Mahmoud, A.E.D., M. Fawzy, and N. Khan, *Artificial Intelligence and Modeling for Water Sustainability: Global Challenges*. 1st Edition ed. 2023: CRC Press.

[54] Y. Cheng and S. P. Jiang, "Advances in electrocatalysts for oxygen evolution reaction of water electrolysis-from metal oxides to carbon nanotubes," *Progress in Natural Science: Materials International*, vol. 25, no. 6. Elsevier B.V., pp. 545–553, Dec. 01, 2015. doi: 10.1016/j.pnsc.2015.11.008

[55] F. Li et al., "Designed synthesis of multi-walled carbon nanotubes@Cu@MoS2 hybrid as advanced electrocatalyst for highly efficient hydrogen evolution reaction," *Journal of Power Sources*, vol. 300, 2015, doi: 10.1016/j.jpowsour.2015.09.084

[56] S. Singla, S. Sharma, S. Basu, N. P. Shetti, and T. M. Aminabhavi, "Photocatalytic water splitting hydrogen production via environmental benign carbon based nanomaterials," 2021, *International Journal of Hydrogen Energy*, doi: 10.1016/j.ijhydene.2021.07.187

[57] L. Clarizia, D. Russo, I. Di Somma, R. Andreozzi, and R. Marotta, "Hydrogen generation through solar photocatalytic processes: A review of the configuration and the properties of effective metal-based semiconductor nanomaterials," *Energies (Basel)*, vol. 10, no. 10, 2017, doi: 10.3390/en10101624

[58] A. V. Puga, "Photocatalytic production of hydrogen from biomass-derived feedstocks," *Coordination Chemistry Reviews*, vol. 315, pp. 1–66, May 2016, doi: 10.1016/j.ccr.2015.12.009

[59] S. Cao and J. Yu, "G-C3N4-based photocatalysts for hydrogen generation," *Journal of Physical Chemistry Letters*, vol. 5, no. 12. 2014. doi: 10.1021/jz500546b

[60] B. Zhao et al., "Direct preparation of hierarchical macroporous β-SiC using SiO2 opal as both template and precursor and its application in water splitting," *Materials Technology*, vol. 31, no. 9, 2016, doi: 10.1080/10667857.2016.1182343

[61] S. G. Kumar, R. Kavitha, and P. M. Nithya, "Tailoring the CdS surface structure for photocatalytic applications," *Journal of Environmental Chemical Engineering*, vol. 8, no. 5. 2020. doi: 10.1016/j.jece.2020.104313

[62] H. Yang, A. L. Meng, L. N. Yang, and Z. J. Li, "Construction of S-scheme heterojunction consisting of Zn0.5Cd0.5S with sulfur vacancies and NixCo1-x(OH)2 for highly efficient photocatalytic H2 evolution," *Chemical Engineering Journal*, vol. 432, 2022, doi: 10.1016/j.cej.2021.134371

[63] Q. Xiang, J. Yu, and M. Jaroniec, "Synergetic effect of MoS 2 and graphene as cocatalysts for enhanced photocatalytic H 2 production activity of TiO 2 nanoparticles," *Journal of the American Chemical Society*, vol. 134, no. 15, pp. 6575–6578, Apr. 2012, doi: 10.1021/JA302846N

[64] C. Li, J. Wang, Z. Jiang, and P. Hu, "Co/Cu2O assisted growth of graphene oxide on carbon nanotubes and its water splitting activities," *New Journal of Chemistry*, vol. 39, no. 6, 2015, doi: 10.1039/c5nj00558b

[65] F. Li, X. Jiang, J. Zhao, and S. Zhang, "Graphene oxide: A promising nanomaterial for energy and environmental applications," *Nano Energy*, vol. 16. 2015. doi: 10.1016/j.nanoen.2015.07.014

[66] K. Wang, S. Liu, L. Zhang, and Z. Jin, "Hierarchically Grown Ni–Mo–S Modified 2D CeO2 for High-Efficiency Photocatalytic Hydrogen Evolution," *Catalysis Letters*, vol. 152, no. 4, 2022, doi: 10.1007/s10562-021-03703-8

[67] G. Mamba and A. K. Mishra, "Graphitic carbon nitride (g-C3N4) nanocomposites: A new and exciting generation of visible light driven photocatalysts for environmental pollution remediation," *Applied Catalysis B: Environmental*, vol. 198. 2016. doi: 10.1016/j.apcatb.2016.05.052

[68] Mahmoud, A.E.D., "Recent advances of TiO2 nanocomposites for photocatalytic degradation of water contaminants and rechargeable sodium ion batteries," in *Advances in Nanocomposite Materials for Environmental and Energy Harvesting Applications*, A.E. Shalan, A.S. Hamdy Makhlouf, and S. Lanceros-Méndez, Editors. 2022, Springer International Publishing: Cham. p. 757–770.

[69] L. Clarizia, D. Russo, I. Di Somma, R. Andreozzi, and R. Marotta, "Hydrogen generation through solar photocatalytic processes: A review of the configuration and the properties of effective metal-based semiconductor nanomaterials," *Energies (Basel)*, vol. 10,
no. 10, 2017, doi: 10.3390/en10101624

[70] R. Kumar, V. A. Suyamburajan, A. Khan, A. M. Asiri, and H. Dzudzevic-Cancar,
"Nanophotocatalysts for hydrogen production applications," *Nanomaterials for Hydrogen
Storage Applications*, pp. 219–229, 2021, doi: 10.1016/b978-0-12-819476-8.00018-9

[71] S. Mahzoon, S. M. Nowee, and M. Haghighi, "Synergetic combination of 1D-2D
g-C3N4 heterojunction nanophotocatalyst for hydrogen production via water splitting
under visible light irradiation," *Renewable Energy*, vol. 127, pp. 433–443, Nov. 2018,
doi: 10.1016/j.renene.2018.04.076

[72] Huang, L., et al., "Enhanced water purification via redox interfaces created by an atomic
layer deposition strategy." *Environmental Science: Nano*, 2021. 8(4): p. 950–959.

[73] P. Ganguly et al., "2D Nanomaterials for Photocatalytic Hydrogen Production," *ACS
Energy Letters*, vol. 4, no. 7, pp. 1687–1709, Jul. 2019, doi: 10.1021/acsenergylett.9b00940

[74] A. Zada, M. Khan, M. N. Qureshi, S. Y. Liu, and R. Wang, "Accelerating photocatalytic
hydrogen production and pollutant degradation by functionalizing g-C3N4 with SnO₂,"
Frontiers in Chemistry, vol. 7, p. 941, Feb. 2020, doi: 10.3389/fchem.2019.00941

[75] S. Balasubramanian, P. Wang, R. D. Schaller, T. Rajh, and E. A. Rozhkova, "High-
performance bioassisted nanophotocatalyst for hydrogen production," *Nano Letters*,
vol. 13, no. 7, pp. 3365–3371, Jul. 2013, doi: 10.1021/nl4016655

[76] X. Li, Y. Hou, Q. Zhao, W. Teng, X. Hu, and G. Chen, "Capability of novel ZnFe2O4
nanotube arrays for visible-light induced degradation of 4-chlorophenol," *Chemosphere*,
vol. 82, no. 4, pp. 581–586, Jan. 2011, doi: 10.1016/j.chemosphere.2010.09.068

[77] P. Pachfule et al., "Diacetylene Functionalized Covalent Organic Framework (COF) for
photocatalytic hydrogen generation," *Journal of the American Chemical Society*, vol.
140, no. 4, pp. 1423–1427, Jan. 2018, doi: 10.1021/jacs.7b11255

[78] P. Kumar, R. Boukherroub, and K. Shankar, "Sunlight-driven water-splitting using two-
dimensional carbon-based semiconductors," *Journal of Materials Chemistry A*, vol. 6,
no. 27. 2018. doi: 10.1039/c8ta02061b

[79] C. Dai, Y. Pan, and B. Liu, "Conjugated polymer nanomaterials for solar water splitting," *Advanced Energy Materials*, vol. 10, no. 42, p. 2002474, Nov. 2020, doi: 10.1002/
aenm.202002474

[80] T. Zhu, H. Bin Wu, Y. Wang, R. Xu, and X. W. Lou, "Formation of 1D hierarchical
structures composed of Ni3S 2 nanosheets on CNTs backbone for supercapacitors and
photocatalytic H2 production," *Advanced Energy Materials*, vol. 2, no. 12, 2012, doi:
10.1002/aenm.201200269

[81] R. K. Chava, J. Y. Do, and M. Kang, "Smart hybridization of Au coupled CdS nanorods
with few layered MoS2 nanosheets for high performance photocatalytic hydrogen evolution reaction," *ACS Sustainable Chemistry & Engineering*, vol. 6, no. 5, 2018, doi:
10.1021/acssuschemeng.8b00249

[82] M. Zhang et al., "High H2 evolution from Quantum Cu(II) nanodot-doped two-
dimensional Ultrathin TiO2 nanosheets with dominant exposed {001} facets for reforming glycerol with multiple electron transport pathways," *Journal of Physical Chemistry
C*, vol. 120, no. 20, 2016, doi: 10.1021/acs.jpcc.6b01030

[83] Q. Xu, B. Zhu, C. Jiang, B. Cheng, and J. Yu, "Constructing 2D/2D Fe2O3/g-C3N4
Direct Z-Scheme Photocatalysts with Enhanced H2 Generation Performance," *Solar
RRL*, vol. 2, no. 3, 2018, doi: 10.1002/solr.201800006

[84] M. V. Pavliuk et al., "Nano-hybrid plasmonic photocatalyst for hydrogen production
at 20% efficiency," *Scientific Reports*, vol. 7, no. 1, pp. 1–9, Aug. 2017, doi: 10.1038/
s41598-017-09261-7

[85] H. Dang et al., "Hydrothermal preparation and characterization of nanostructured CNTs/ ZnFe 2 O 4 composites for solar water splitting application," *Ceramics International*, 2016, doi: 10.1016/j.ceramint.2016.03.019

[86] M. Mamathakumari, D. Praveen Kumar, P. Haridoss, V. Durgakumari, and M. V. Shankar, "Nanohybrid of titania/carbon nanotubes – Nanohorns: A promising photocatalyst for enhanced hydrogen production under solar irradiation," *International Journal of Hydrogen Energy*, vol. 40, no. 4, 2015, doi: 10.1016/j.ijhydene.2014.11.117

[87] H. S. Sajjadizadeh, E. K. Goharshadi, and H. Ahmadzadeh, "Photoelectrochemical water splitting by engineered multilayer TiO2/GQDs photoanode with cascade charge transfer structure," *International Journal of Hydrogen Energy*, vol. 45, no. 1, 2020, doi: 10.1016/j.ijhydene.2019.10.161

[88] N. Lakshmana Reddy, K. K. Cheralathan, V. Durga Kumari, B. Neppolian, and S. Muthukonda Venkatakrishnan, "Photocatalytic reforming of biomass derived crude glycerol in water: A sustainable approach for improved hydrogen generation using Ni(OH)2 decorated TiO2 nanotubes under solar light irradiation," *ACS Sustainable Chemistry & Engineering*, vol. 6, no. 3, pp. 3754–3764, Mar. 2018, doi: 10.1021/ ACSSUSCHEMENG.7B04118

[89] X. Han et al., "WO3/g-C3N4 two-dimensional composites for visible-light driven photocatalytic hydrogen production," *International Journal of Hydrogen Energy*, vol. 43, no. 10, 2018, doi: 10.1016/j.ijhydene.2018.01.117

6 Modeling and Optimization of Nanomaterials Production Processes

Lanrewaju I. Fajimi, Bilainu O. Oboirien, and Zainab T. Yaqub

6.1 INTRODUCTION

Nanotechnology (NT) applications are predicted to have an impact on almost every part of life and enable significant advancements in materials science, health, manufacturing, and knowledge-based technologies amongst several others [1]. A key fundamental component of NT is nanomaterial (NMT). According to ISO/TS 80004 standard, NMT is a material with both interior and exterior dimensions in the nanoscale (i.e., having an internal structure or surface structure in the nanoscale, with a length range from 1 nm to 100 nm [2, 3]. This incorporates both discrete NMTs (nano-objects) and nanostructured materials. Huge efforts have been targeted at the production of fruitful NMTs, nevertheless, there are still several factors that needed to be investigated [4].

Currently, NMTs are playing a bigger role in the advancement of humanity as a whole. For instance, employing green technology that relies solely on NT is the only way to prevent global warming and climate change in the first place [5] due to the fact that it has been demonstrated that bulk material-based technology cannot compete with nanotechnology in terms of effectiveness [5, 6]. Furthermore, tools to diagnose and manage the epidemic diseases that are raging across the globe are being developed using NMTs. For instance, in 2019, nanoparticles allowed for the diagnosis and management of Covid-19, a disease that was killing a lot of people worldwide [7, 8]. Also, NMTs could be used to identify and cure the monkey pox disease, which is now raging around the world [9]. Nanoscience and NT are predicted to have a major impact on how the world develops in the future [5, 9]. Generally, NMTs are divided into five main categories and several subcategories (see Figure 6.1). The main categories are size (0–3 dimensions), toxicity (fiber-like, bio-persistent granular and CMAR), architectural configuration (composite, organic, inorganic, and carbon-based), origin (artificial and natural), and size of their pores (micropores, mesopores, and macropores) [5].

Several approaches have been used in NMT production however, chemical vapor deposition (CVD) stands tall. This is due to its ability to accurately control NMT properties such as length, diameter, purity, and density, amongst several others. Due to its low cost and ease of use, the CVD process is frequently both theoretically and

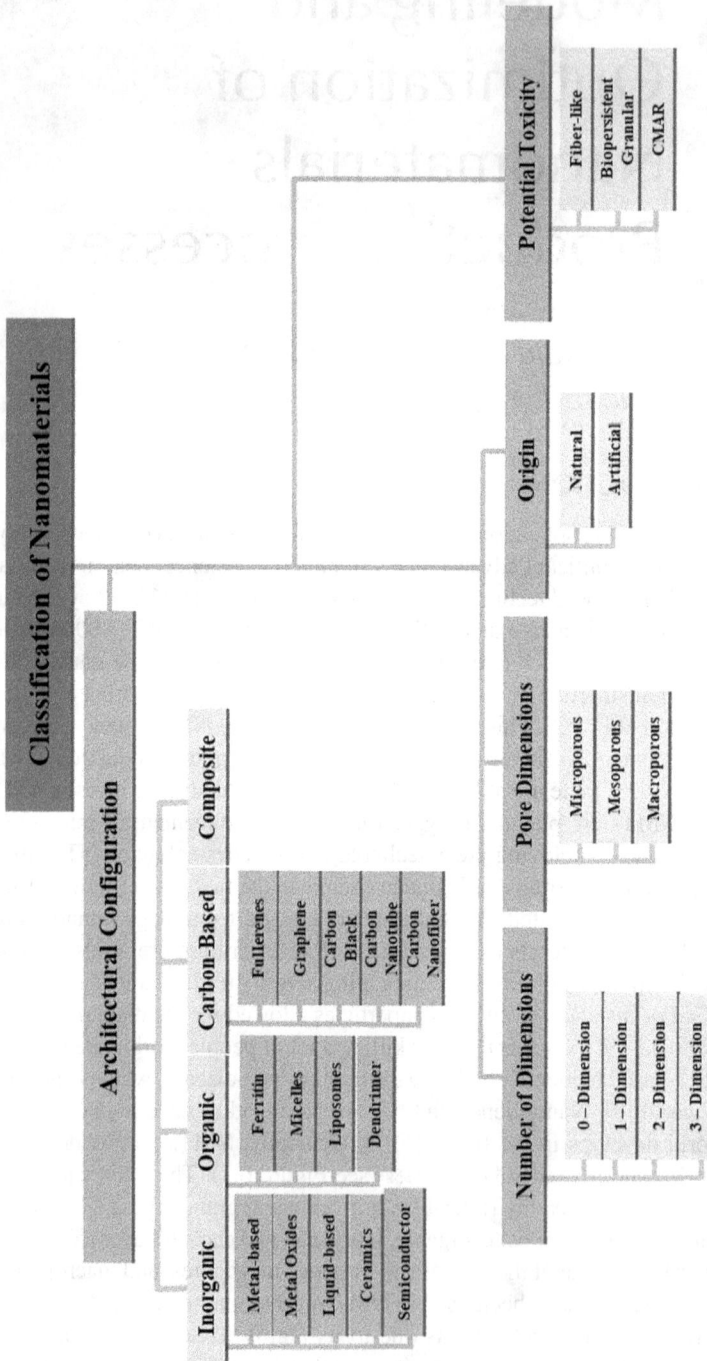

FIGURE 6.1   Classification of nanomaterials.

empirically examined [4, 10]. Production of NMT via CVD is accomplished by using a carbon source in the gas phase (precursor) and an energy source, such as a plasma or a resistively heated coil, to transfer energy into the gaseous carbon molecule [11]. Even though NMT production through CVD is widely preferred, it has some drawbacks which are mainly due to the high toxicity of the gaseous byproducts and the need for specialized equipment [12].

In order to get the most performance out of NMTs, a variety of additional factors, like the NMT tube curvature, have been proven to be important. However, experimental outcomes have consistently fallen short of the rule of mixtures expectations regardless of the type of nanotube influence on various attributes [13, 14]. In addition, to model and optimize NMT production, the simultaneous control of the size and shape of the particles along with their uniformity is crucial [15]. Hence, there is a need to re-investigate NMT production from a modeling point of view. Computational modeling and optimization tools have always played a significant role in the revolution of NMT science [14, 16].

Investigating various models that could be used to maximize NMT, for example, carbon nanotube (CNT) production, is necessary to overcome the challenges associated with studying the behavior of NMTs, such as the expense of laboratory experiments or other conventional methods that require a lot of time and resources. The objective of this chapter is to investigate different models and optimization tools that have been employed in CNT production processes.

6.2 ARCHITECTURAL/STRUCTURAL NANOMATERIALS

As aforementioned, NMTs have subcategories, however, this section will focus on the architectural/structural subcategories. These NMT subcategories include composite, organic, inorganic, and carbon-based. The distinction between these four as well as different applications will be discussed briefly in subsequent sections.

6.2.1 COMPOSITE NMTs

These NMTs consist of nanoparticles coupled with other nanoparticles, nanoparticles mixed with large-scale materials, and NMTs combined with bulk-type materials [17]. Automobile parts and packaging materials, among other things, are already using composite NMTs to enhance their mechanical, thermal, and flame-retardant qualities [5].

6.2.2 ORGANIC NMTs

Organic compounds are transformed into organic nanomaterials at the nanoscale. Ferritin, liposomes, dendrimers, micelles, and organic nanoparticles or polymers are a few examples. Liposomes and nano-capsule micelles are biodegradable, non-toxic nanoparticles with hollow interiors that are sensitive to heat, electromagnetic radiation, and light [18]. A dendrimer's surface is covered in multiple chain endings that can carry out particular chemical processes [5, 18]. In molecular recognition, nano-sensing, light harvesting, and opto-electrochemical systems, dendrimers are employed [5]. In addition, three-dimensional (3D) dendrimers may be helpful for drug delivery since they have interior holes that can contain additional molecules [18, 19].

6.2.3 INORGANIC NMTS

Inorganic NMTs are also divided into six classes: Metals, metal oxides, quantum dots, metal\CNT hybrids, metal oxides\CNT hybrids, and quantum dots\CNT hybrids. Metallic NMTs are defined as nanoscale metals with dimensions between 1 and 100 nm [20]. The most intriguing metals being examined are gold, silver, palladium, platinum, zinc, cadmium, copper, and iron. They display several traits that give them unique NMT qualities [15]. Due to their distinctive chemical, magnetic, electrical, mechanical, and optical properties, metal oxides like aluminum oxide (Al_2O_3), cerium oxide (CeO_2), copper oxide (CuO), iron(iii)oxide (Fe_2O_3), nickel(ii)oxide (NiO), titanium(iv)oxide (TiO_2), zinc oxide (ZnO), zirconium dioxide (ZrO_2), amongst several others, have attracted a lot of study attention [15, 21–25] and have a wide variety of specific applications in numerous technical domains [26, 27].

Another category of inorganic NMTs are quantum dots (QDs). QDs are monodisperse crystalline clusters that have physical dimensions that are lower than the bulk-exciton Bohr radius [28]. A variety of these dots have been constructed, often made of atoms from groups II-VI, III-V, or IV-VI. Furthermore, QDs may include NMTs such as CdS, CdSe, CdTe, ZnSe, CdSe/ZnS, In, amongst several others [15], and have diameters in the range of 2–10 nanometers. To create NMTs, metal oxides, and hydroxides have been successfully placed on CNTs. The resulting NMTs combine the distinctive qualities and capabilities of the two types of components and may also display certain novel features brought on by the cooperative effects between the two types of materials [29]. Notably, CNTs with a lot of chemically active surface area are excellent supports for QDs. An intermediary molecule, such as a polymer that has already been conjugated to either the CNT or the QD, is required for the indirect bonding [30–32].

6.2.4 CARBON-BASED NMTS

Carbon-based NMTs include graphene, carbon nanotube (CNT), carbon nanofiber (CNF), and fullerenes. Figure 6.2 shows the structural model of different carbon-based NMTs. Graphene is an intriguing substance with exceptional qualities like extraordinary transparency, ultra-high electrical and thermal conductivities, and extreme mechanical strength [33–35]. In graphene, carbon atoms are arranged in a honeycomb crystal lattice to form a one-atom-thick planar sheet [1, 36–38]. Unlike graphene, CNT is a cylindrical-shaped molecule with a single or complex roll(s) of graphene sheet(s). Fullerenes on the other hand are hollow cage-like structures. In some cases, graphene and CNTs are classified under the fullerene family [1]. The last example of carbon-based NMTs is CNF. Just like CNT, CNF is also cylindrical, however, the graphene layers in CNF are arranged as stacked plates (or cones). CNT is also referred to as CNF with graphene layers wrapped into perfect cylinders [39]. CNFs are less affected by van der Waals forces and tend to stay scattered for longer periods, in contrast to CNTs, which are affected by the forces and induce the nanotubes to form ropes or reassemble after being dispersed [39]. From the various sub-categories of NMTs, this chapter will focus on CNT production modeling and optimization.

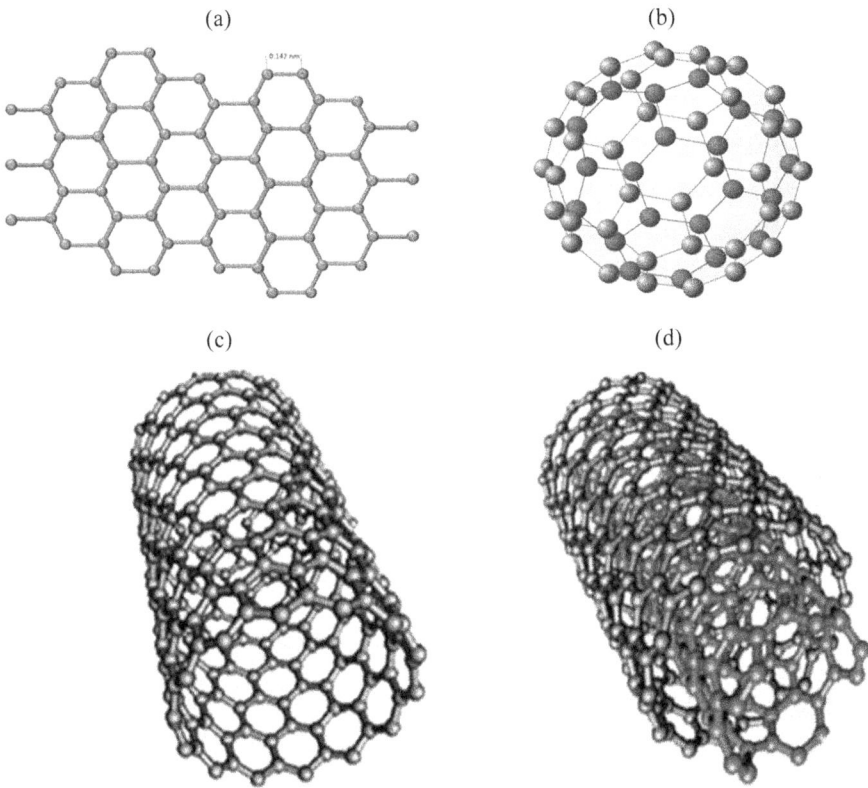

FIGURE 6.2 Schematic representation of carbon-based nanomaterials architecture. (a) Graphene [40] (b) Buckminsterfullerene (C_{60}) [41] (c) SWCNT [42] (d) MWCNT [42].

(a, c & d Reprinted with permission from [40, 42]).

6.3 CARBON NANOTUBES

Carbon nanotubes (CNTs) are hollow cylindrical tubes made up of carbon with nano-sized diameters within the 0.1–100 nm and long lengths above 100 μm [43, 44]. They have a high electrical conductivity and tensile strength that is more than 100 times greater than that of stainless steel, and they are physically and chemically stable [43]. Moreover, CNTs have a wide range of uses, including energy storage, biosensors, medical devices, conductive paints and coatings, semiconductor transistors for environmental applications, microelectronics, and composite plastic materials as electrically conductive fillers or to increase composite strength [43, 45–47]. In the earliest attempts to produce CNTs, high-temperature preparation techniques were used [11, 48, 49], like the arc discharge approach or the laser ablation technique. Currently, CNTs are being created utilizing a variety of processes, including CVD, flame synthesis, sonochemistry, ball milling, and others because of technological advancements in the field [11]. Arc discharge, laser ablation, electrolysis, and CVD are a few of the techniques (see Figure 6.3) that are frequently applied in industry. Further, CNTs can either be single-walled or multi-walled [11, 50]. Although single-walled

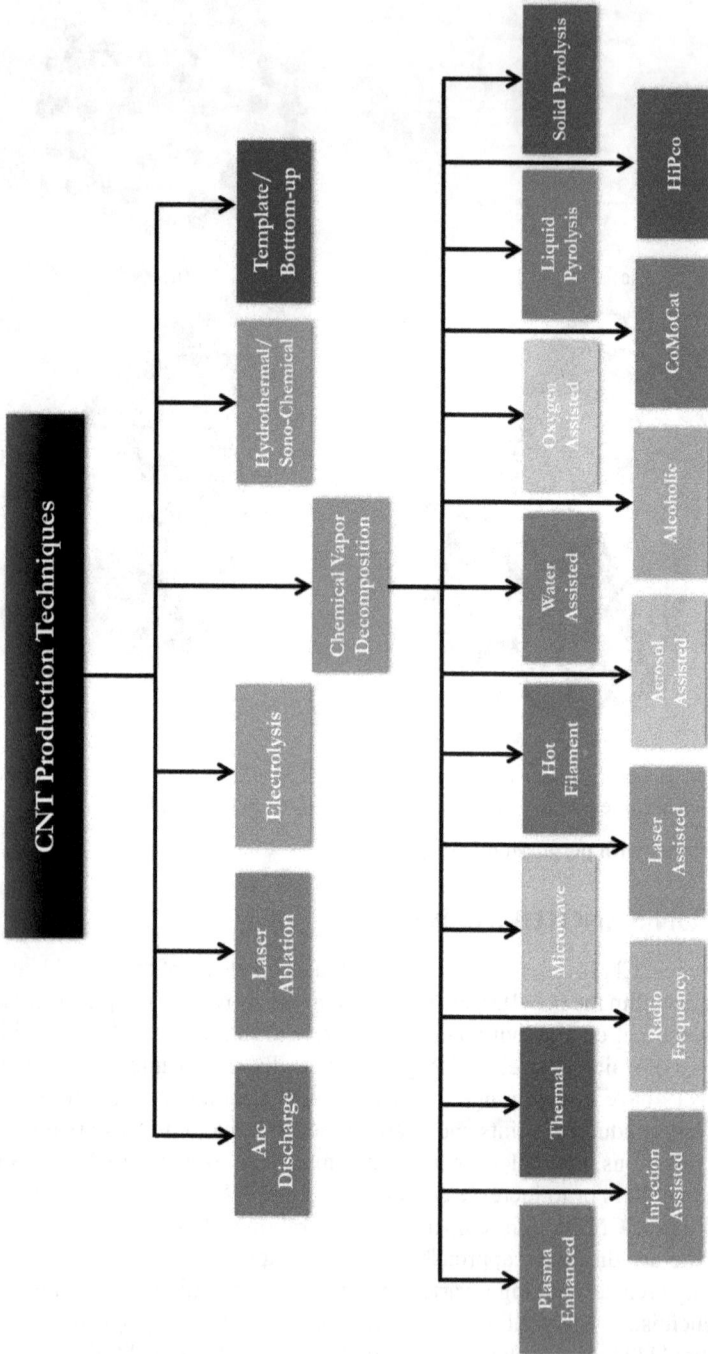

FIGURE 6.3 CNT production techniques

(Image recreated with permission from reference [75]).

CNTs (SWCNTs) were initially found, multi-walled CNTs (MWCNTs) have not received as much attention [44]. This is due to the higher specific stiffness and strength of SWCNTs when compared to MWCNTs. However, MWCNTs provide better advantages such as having lower resistance, lower power consumption, higher mechanical strength, higher thermal stability, higher conductivity, etc. than SWCNTs [51, 52]. A range of different carbon sources for the production of CNTs has been investigated, including coal [53], plastic tar [54–56], tires [57–59], methane [60–62], carbon dioxide [62, 63], ethanol [49], and aromatic hydrocarbons including benzene, toluene, and xylene [43, 64–66] and several other sources [43, 67–70]. Carbon nanotubes are a significant kind of petrochemical raw material that is extensively employed in the separation, refinement, and processing of chemical industry products like petrochemical, gasoline, and natural gas [71–74].

6.3.1 Carbon Nanotube Production Modeling

Modeling CNT production is a huge task for most researchers; however, some authors have managed to overcome this using different models which include the kinetic model and its derivatives, process models such as the fused deposition model, as well as machine learning models to predict critical CNT parameters. The sections below describe some of these models and how they have been applied in CNT modeling.

6.3.1.1 Modeling of Ethanol to CNT

Carbon nanotubes have been synthesized in different reactors, using different techniques and carbon precursors. In this case study, the authors employed a fluidized bed reactor using the chemical vapor deposition (CVD) of pure ethanol which serves as a carbon precursor. The experiment was carried out in the presence of a particulate catalyst made of iron and cobalt nanoparticles supported on alumina [49]. The furnace encircling the reactor was able to regulate the temperature within. As a result, the process can be thought of as being isothermal, and the fixed point chosen for the furnace can be considered the process temperature. The isothermal state of the reactor is crucial both for achieving high productivity and for preventing the development of amorphous carbon [49, 76]. The CNT synthesis overall reaction is given by Equation (6.1):

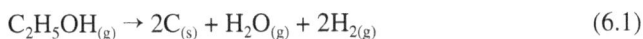

$$C_2H_5OH_{(g)} \rightarrow 2C_{(s)} + H_2O_{(g)} + 2H_{2(g)} \qquad (6.1)$$

According to [49], the first-order reaction rate which corresponds to EtOH cracking is defined as:

$$r_{EtOH} = -k_{EtOH}(C_{EtOH})D_p \qquad (6.2)$$

where r_{EtOH}, C_{EtOH}, and $-k_{EtOH}$ are ethanol reaction rate (mol/m³s), concentration (mol/m³), and reaction constant (1/s) respectively and D_p implies dense phase. $-k_{EtOH}$ according to Arrhenius's law is a function of temperature (T) and activation energy (E_a) as shown in Equation (6.3).

$$-k_{EtOH} = k_0 e^{\frac{E_a}{RT}} \qquad (6.3)$$

According to the established mechanism for CNT synthesis, the catalyst dissolves the carbon that has been broken down and deposited on the active sites (nano-particles of transition metals) [49]. Carbon precipitates out as CNTs when the point of carbon saturation is achieved. It is plausible to believe that carbon is deposited in the form of CNTs and that the creation of CNTs on the catalyst is the limiting stage since ethanol degradation occurs in the presence of catalytic nano-particles. Since CNTs are deposited on active sites, the catalyst activity (φ) – a term used to describe how many active sites are present on the catalyst – directly affects how many CNTs are formed. As a result, the intrinsic reaction additionally has to include catalyst activity. Hence, Equations (6.3) and (6.2) with φ become Equations (6.4) and (6.5):

$$-k_{\text{EtOH}} = k_0 \varphi e^{-\frac{E_a}{RT}} \tag{6.4}$$

$$r'_{\text{EtOH}} = -k_{\text{EtOH}} \varphi \left(C_{\text{EtOH}} \right)_{D_p} \tag{6.5}$$

When the active site on the catalyst is covered by a single nanotube, φ becomes:

$$\varphi = \frac{\alpha_0 - \gamma}{\alpha_0} \tag{6.6}$$

where α_0 is the initial number of active sites and γ is the number of active sites covered with CNT. For fresh catalysts, since none of the active sites are filled with nanotubes, $\gamma = 0$, hence, $\varphi = 1$. In addition, the total CNT formed varies with time. As nanotubes formation progresses on the sites, $\varphi \to 0$ at the end of the reaction hence, $\gamma = \alpha_0$. Hence, Equation (6.6) can be written as:

$$\frac{d\varphi}{dt} = \left(-\frac{1}{\alpha_0} \right) \left(\frac{d\gamma}{dt} \right) \tag{6.7}$$

Equation (6.7) expresses the catalyst deactivation rate which depends on the carbon deposition rate, which also relates to the ethanol consumption rate. In order to determine the catalyst deactivation rate, α_0 must be known and this can only be done using reasonable approximations which correspond to the composition of the catalyst.

Modeling of CNT production in a fluidized bed reactor is divided into two phases: the bubble phase and the dense phase. While the bubble phase containing the gas phase is seen as plug flow, the dense phase comprising solid catalysts and a cloud of gas phase is regarded as a perfect mixing condition (as a pseudo-homogeneous phase). The gas reactant and gas product are both contained in the bubble phase's inlet and outflow, respectively. Since mass and heat transmission connects these two stages, the process model is dependent on both space and time. According to [49], the dense phase mass balance is given in Equation (6.8) as:

$$q_{D_p}\left(C_{EtOH}\right)_F - q_{D_p}\left(C_{EtOH}\right)_{D_p} + \int_0^H \left(K_{bd}\right)\delta A\left(\left(C_{EtOH}\right)_{B_p} - \left(C_{EtOH}\right)_{D_p}\right)dy$$

$$-\left(1-\delta\right)\left(1-\varepsilon_d r'_{EtOH}AH\right) = 0 \qquad (6.8)$$

where qD_p is the volumetric flow rate (m³/s) in the dense phase, K_{bd} is the overall mass transfer coefficient between the two-phase (1/s) δ is the fractional volume of bubbles, A is the reactor cross-sectional area (m²), ε_d is the dense phase voidage, H is the height of the reactor (m).

The bubble phase mass balance is given in Equation (6.9) as:

$$q_{B_p}\left(\frac{d\left(C_{EtOH}\right)_B}{dy}\right) = -\left(K_{bd}\right)\delta A\left(\left(C_{EtOH}\right)_{B_p} - \left(C_{EtOH}\right)_{D_p}\right) \qquad (6.9)$$

where qB_p is the volumetric flow rate (m³/s) in the bubble phase.

The molar rate of ethanol converted to carbon can be expressed in Equation (6.10) as:

$$n_{EtOH} = \left(1-\delta\right)\left(1-\varepsilon_d\right)r'_{EtOH}AH \qquad (6.10)$$

The mole of carbon formed in the reaction is $n_C = n_{EtOH}$ as the stoichiometric of the reaction (Equation 6.1). Then, the rate for the number of carbon atoms formed can be expressed as in Equation (6.11):

$$Nu_{C_{atm}} = n_C \times N_{Avo} = 2N_{Avo}\left(1-\delta\right)\left(1-\varepsilon_d\right)r'_{EtOH}AH) \qquad (6.11)$$

By using a software called "wrapping" (available at: www.photon.t.u-tokyo.ac.jp), one can estimate the number of carbon atoms present in each CNT. The mean value of the diameter of the CNT can also be estimated using electron microscopy observations [75]. Equation (6.5) predicts that the rate of CNT creation (NuCNT) will be as follows:

$$Nu_{CNT} = \frac{Nu_{C_{atm}}}{N_{C-CNT}} = \left(\frac{dy}{dt}\right) \qquad (6.12)$$

Combining Equations 6.5, 6.7, 6.11, 6.12, the general equation can be expressed as Equation 6.13):

$$\frac{d\varphi}{dt} = \frac{2N_{Avo}}{N_{C-CNT}}\left(-\frac{1}{\alpha_0}\right)\left(1-\delta\right)\left(1-\varepsilon_d\right)AHk_0\varphi e^{-\frac{E}{RT}}\left(C_{EtOH}\right) \qquad (6.13)$$

6.3.1.2 Kinetic-based Derivative Models Employed in CNT Production Modeling

Various models have been employed in modeling CNT production process, some of these processes are presented in Table 6.1. From a survey of some research works

TABLE 6.1
Models Employed in CNT Production and Key Observations

SN	Model Name	Work	Model and Key CNT Reaction	Key Observation(s)	Ref.
1	Simple model based on weight	Effect of temperature on the kinetics of C_2H_4 decomposition over reduced iron oxide catalyst for CNT production	$$CNT_{yield}\,(\%) = \left[\frac{W_{cat.CNT} - (W_{cat0} - W_{catloss})}{(W_{cat0} - W_{catloss})} \right] \times 100$$ where $W_{cat,CNT}$, W_{cat0}, $W_{catloss}$ are the weight (g) of the CNT and catalyst deposited, the initial weight of the catalyst, and the weight of the loss of catalyst at the operating condition respectively.	The result from the proposed model shows that the percentage yield of CNT produced is significantly influenced by the iron oxide crystal size and decomposition temperature.	[67]
2	Weight gain Model	Synthesis of carbon nanotubes from conventional biomass-based gasification gas	$CH_{4(g)} \rightarrow C_{CNT(s)} + 2H_{2(g)}$ $2CO_{(g)} \rightarrow C_{CNT(s)} + CO_{2(g)}$ $CO_{(g)} + H_{2(g)} \rightarrow C_{CNT(s)} + H_2O_{(g)}$ $$\frac{C_{dep}}{C_{cat}} = \frac{W_t - W_{cat}}{W_{cat}} \times 100$$ where $\frac{C_{dep}}{C_{cat}}$ is the carbon weight gain per unit catalyst weight, W_t and W_{cat} are the total and initial weights (g) of the catalyst after the reactions.	This model followed the assumption that CNT was produced from two carbon sources (CH_4 and CO). The authors observed that the CNT produced is a function of the initial and total weight of the catalyst.	[62]
3	Kinetic Model	Mechanism and kinetics of growth termination in controlled CVD growth of MWCNT arrays.	$C_2H_{4(g)} \rightarrow 2C_{CNT(s)} + 2H_{2(g)}$ $$v_{gCNT} \sim v e^{\frac{E_a}{k_BT}} KP_{H_2}^2 \left(\frac{P_{C_2H_4}}{P_{H_2}^2}\frac{1}{K} - 1 \right) = v e^{\frac{E_a}{k_BT}} \left(\xi_{C_2H_4} P_0 - K \xi_{H_2}^2 P_0^2 \right)$$ where v_{gCNT} – CNT growth rate, v – pre-exponential frequency factor, E_a – activation barrier, k_B – Boltzmann constant (j/K), T – Temperature (K), K is the equilibrium constant, ξi – volume fractions of gaseous specie, P_0, P_i are the total pressure and partial pressures of gaseous specie (bar).	In addition to describing the growth rate as a function of total process pressure, the model also incorporated the effects of partial pressures of C_2H_4 and hydrogen on C_2H_4 decomposition to CNT.	[68]

| 4 | Kato and Wen Model combined with Kinetic Model | Kinetic modeling study of CNT synthesis by fluidized bed CVD | $C_2H_{4(g)} \rightarrow 2C_{CNT(s)} + 2H_{2(g)}$

$\left(\dfrac{dX_{CNT}}{dt}\right) = 69.97 \times \left(\%_{Fe} \times m_{cat}\right)^{0.28} \times e^{\frac{29000}{RT_K}} \times \left(y_{C_2H_4}\right)^{\frac{3}{4}}$

where $\left(\dfrac{dX_{CNT}}{dt}\right)$ – CNT formation rate,

$\%_{Fe}$ – composition of Fe in the catalyst (wt.%),
m_{cat} – mass of the catalyst (kg),
R – gas constant (J/mol K),
T_K – bed temperature (K),
y_{C2H4} – molar composition of C_2H_4 (mol%). | The influence of the primary operational parameters on the temporal evolutions of the species molar fractions, the weight of CNTs produced, and the bed properties were numerically investigated in MWCNT production. The proposed model was developed from experimental correlation. | [77] |
| 5 | Kinetic Model | A chemical kinetic model for CVD of CNT | $C_2H_{2(g)} \rightarrow 2C_{(g)} + H_{2(g)}$
$2C_{(g)} + 6Fe_{(s)} \rightarrow 2Fe_3C_{(l)}$
$2Fe_3C_{(l)} \rightarrow 3Fe_{2(s)} + 2C_{CNT(s)}$

$n = \dfrac{N_A \times \rho}{C_{A0}(1 - X_A) \times M_c}$

where n – number of carbon atoms in the CNT produced,
N_A – Avogadro number;
ρ – density Fe_3C (g/m³),
C_{A0} – initial composition of Fe_3C (mol/m³),
X_A – molar conversion of C.
M_c – molar mass of C (g). | The result obtained from the model shows deposition temperature and deposition pressure exhibit a decreasing trend with an increase in furnace temperature up to 1000 K. In addition, as the reaction time increases, the deposition temperature decreases. | [78] |

(Continued)

TABLE 6.1 (Continued)
Models Employed in CNT Production and Key Observations

SN	Model Name	Work	Model and Key CNT Reaction	Key Observation(s)	Ref.
6	Kinetics model for CNT growth	Temperature and time dependence study of SWCNT growth by CVD	$C_2H_5OH_{(g)} + 2S_v \xrightarrow{k_1} 2C.S_{(s)} + H_2O_{(g)} + 2H_{2(g)}$ $C^g_{SWCNT}.C^d_{AMOR}.S + C.S \xrightarrow{k_2} C^{g+1}_{SWCNT}.C^d_{AMOR}.S + S_v$ $C^g_{SWCNT}.C^d_{AMOR}.S + C.S \xrightarrow{k_3} C^g_{SWCNT}.C^{d+1}_{AMOR}.S + S_v$ $C^g_{SWCNT}.C^d_{AMOR}.S + C.S \xrightarrow{k_4} C^{g-1}_{SWCNT}.C^d_{AMOR}.S + H_2 + CO$ $C^g_{SWCNT}.C^d_{AMOR}.S + H_2O \xrightarrow{k_4} C^g_{SWCNT}.C^{d-1}_{AMOR}.S + H_2 + CO$ $\dfrac{d[C.S]}{dt} = 2.k_1 \left[C_2O_5OH \right] \left[S_v \right]^2 - k_2 \left[C.S \right]^2$ $\quad - k_3 \left[C.S \right] \left(\left[C.S \right] + \left[C_{SWCNT} \right] + \left[C_{AMOR} \right] \right)$ where S_v is the catalyst site, g and d, are the numbers of SWCNT and amorphous carbons present respectively.	This study discussed the existence of a temperature window and the best period for growth. The authors proposed that the conflict between the growth of SWCNT and that of amorphous carbon can be used to explain temperature and time dependence.	[79]
7	An improved model for CNT growth	Growth of CNTs in a floating catalyst reactor	$C_8H_{10(g)} \rightarrow 8C_{CNT(s)} + 5H_{2(g)}$ $G_{CNT} = \dfrac{\displaystyle\int_0^t 8CAU_r \left(\dfrac{D_{in}}{D_{in}+D_{out}} \right) dt}{\dfrac{\pi}{4} \dfrac{1}{M_c} nD^2_{CNT} \rho_{CNT} A_s}$ where G_{CNT} is the CNT growth (m), C is the molar composition of CNT (mol/m^3), A is the surface area of the reactor inlet, U_r is the velocity of feedstock in the reactor (m/s), D_{in} and D_{out} is the diameter of the substrate and reactor, M_c is the molar mass of C_{CNT} (g/mol), n is the number of nanotubes per unit surface area (1/m^2), D_{CNT} is the diameter of the CNT, ρ_{CNT} is the CNT density (g/m^3), and A_s is the substrate cross-sectional surface area.	The observation from the developed model shows that catalyst deactivation takes place kinetically when the formation rate of the graphene network is smaller than the carbon deposition rate by the decomposition of CH_4.	[64]

8.	Modified Kinetic Model	Mathematical Modeling of CNT Formation in Fluidized Bed CVD	$C_2H_5OH_{(g)} \rightarrow 2C_{(s)} + H_2O_{(g)} + 2H_{2(g)}$	The result from the proposed model shows that catalyst loading, reaction time, and temperature have a significant impact on how quickly the reaction proceeds and, consequently, how much ethanol is present in the exit flow. The mass of CNTs that develop on the catalyst is also influenced by these variables.	[49]

$$\frac{d\varphi}{dt} = \frac{2N_{Avo}}{N_{C-CNT}}\left(-\frac{1}{\alpha_0}\right)(1-\delta)(1-\varepsilon_d)AHk_0\varphi e^{\frac{E}{RT}}(C_{Eth})$$

where φ is the catalyst activity parameter, N_{Avo} is the Avogadro Number, N_{C-CNT} Number of carbon atoms involved in CNT formation, α_0 is the initial number of active sites, δ is the fractional volume of bubbles, ε_d is the dense-phase voidage, A is the reactor cross sectional area (m²), H is the height of the active part of the reactor (m), k_0 is the preexponential kinetic constant (1/s), E is the activation energy (kJ/mol), R is the gas constant (J/mol K), T is the temperature (K), and C_{Eth} is the ethanol concentration (mol/m³).

[49, 62, 64, 67, 68, 77–79], the kinetic model (and its derivatives) is the most widely employed in CNT production modeling. Table 6.1 also presents the key carbon source, the key reaction(s), as well as key observations from the models that have been employed in CNT production process.

6.3.1.3 Machine Learning Models Employed in CNT Production Modeling

With the use of experience and automation, a system can learn and develop through the use of machine learning, an application of artificial intelligence [80–82]. By developing algorithms that can take inputs and use statistical analysis to predict outputs, such systems may reliably anticipate the outputs [83, 84]. Additionally, upon obtaining fresh data, the outputs are updated automatically. Artificial Neural Networks (ANN), Gaussian Process Regression (GPR), Decision Tree Regression (DTR), Prey Predator Algorithm, Linear Regression (LR), Support Vector Machine Regression (SVMR), Density Functional Theory (DFT), and K-nearest neighbors regression (KNNR) are just a few of the machine learning algorithms that have attracted a lot of attention for their ability to predict CNT properties [85–90]. A summary of some of these models is presented in Table 6.2.

TABLE 6.2

Summary of Some Machine Learning Models Employed in Modeling CNT Production

SN	Work	Key Observation(s)	Ref
1	Machine learning methods for MWCNT genotoxicity prediction	In order to accurately forecast the genotoxicity of various types of MWCNTs, the authors constructed a predictive nano-informatics model that is thoroughly validated and based on statistical and machine learning methodologies. A variety of computational workflows which combined unsupervised (Principal Component Analysis, PCA) and supervised classification methods (Support Vector Machine, "SVM," Random Forest, "RF," Logistic Regression, "LR," and Nave Bayes, "NB") with Bayesian optimization were all employed. The most crucial variables were chosen using the Recursive Feature Elimination (RFE) approach. From their results, RF was the most accurate model for predicting the genotoxicity of MWCNTs with 80% external validation accuracy and also has a high classification probability. The authors concluded that the length, zeta average, and purity were observed to be the most informative MWCNT features in the process.	[86]

(Continued)

TABLE 6.1 (Continued)

Summary of Some Machine Learning Models Employed in Modeling CNT Production

SN	Work	Key Observation(s)	Ref
2	Machine Learning-Assisted High-Throughput Molecular Dynamics Simulation of High-Mechanical Performance CNT	The authors explored chemical detection using cross-sensitive CNT-based field-effect transistors (FET) sensor arrays made of SWCNTs coated with metal nanoparticles. Five purine compounds – adenine, guanine, xanthine, uric acid, and caffeine – were distinguished by integrating the investigation of FET device properties with supervised learning techniques. DFT computations were used to support the interactions between purine compounds and SWCNTs coated with metal nanoparticles. From the results obtained, the method successfully detects the presence of caffeine in 95% of the samples with 48 features using linear discriminant analysis (LDA), and in 93.4% of the samples with only 11 features using SVM approach. The authors concluded that the main advantage of the LDA model is the 95% prediction accuracy.	[91]
3	Development of artificial intelligence-based model for the prediction of Young's modulus of polymer/CNT composites	The authors employed an ANN model to predict the behavior of composites made of polymers and carbon nanotubes (CNTs). A parametric analysis was used to choose and refine the design of the ANN model. The obtained results demonstrated that the suggested model performs admirably during both the training and testing stages, with correlation coefficients of 0.978 and 0.986, respectively.	[87]
4.	Prediction of tensile strength of polymer CNT composites using practical machine learning method	The authors developed a machine learning-based model that can be used to predict the tensile strength of composites made of polymer and CNTs. A database with 11 input variables was constructed using literature data. The primary factors considered for predicting the tensile strength of nanocomposites include polymer matrix type, polymer matrix mechanical properties, CNT physical characteristics, and CNT mechanical properties amongst several others. Gaussian Process Regression (GPR) model was employed in the prediction process. The authors concluded that the GPR model performed well in both the training and testing phases (RMSE = 5.982 and 5.327 MPa, MAE = 3.447 and 3.539 MPa, respectively).	[88]
5.	CNT Growth Rate Regression using SVM and ANN Machines and Artificial Neural Net	The author employed an SVM classifier to train an SWNT dataset which comprise 121 experiments out of a total of 450 experiments. With an accuracy of 95.04%, the SVM classifier predicted parameter values for SWNT growth with 99% probability. In addition, the author also employed ANN to forecast SWNT growth rates and growth lengths. From the result obtained, a peak in growth rate as a function of water concentration and growth temperature was shown in the analysis of the ANN growth rate model. The growth length model demonstrated an 80% reduction in validation errors and was trained using the same growth experiments as the growth rate model. The author concluded that The SWNT length-maximizing water/ethylene ratio was likewise determined by the growth length model.	[89]

6.3.1.4   Other Models Employed in CNT Production Modeling

Besides the kinetic-based model and its derivatives, and the machine learning model, other models that have been employed in modeling CNT production include:

- **Fused Deposition Model (FDM)**: The most important additive manu-facturing approach is FDM, which describes the process of depositing successive layers of material in a computer-controlled environment to produce a 3D object such as CNT. The limited options of materials that are now accessible and the fact that parts created using FDM method are only employed as conceptual or demonstration parts rather than functional parts are the main drawbacks of employing FDM in industrial applica-tions [14].
- **Coarse Grain Model (CGM)**: To transfer atomistic systems onto systems with a coarser resolution, CGM techniques like Monte Carlo simulations are developed. Consequently, CGM modeling approaches can be used to simulate larger systems with longer time scales and less processing [92].
- **Molecular Dynamic Model (MDM)**: MDM is one of the most effective modeling CNT techniques because it mimics nanoscale systems using the laws of physics governing particles and motion. The biggest drawback of MDM is that a variety of simulation approaches can be used to make mod-els simpler without compromising the accuracy of the results [93].
- **Density Function Theory (DFT)**: Based on quantum physics, DFT inves-tigates the electrical structure of atomic-scale systems. The local density approximation (LDA), which solely takes into account the electronic den-sity at each place in space, is one of the approximations that the DFT models frequently employ. Although DFT models simulate molecular structure and provide the stable state's geometric characteristics through energy minimi-zation, these models have some drawbacks. It is demonstrated, for instance, that LDA is unable to faithfully recreate systems with van der Waals interac-tions and weak bonds [94].

6.3.1.5   Thermochemical Processes and Reactors Employed in CNT Production Modeling

Various thermochemical processes have been employed in CNT production, how-ever, most of these techniques are employed to break down the main carbon source waste materials (such as plastics, coal, and tires) into the desired carbon source for CNT production. Thermal cracking, pyrolysis, and gasification are the common thermochemical processes that precede chemical vapor deposition (CVD) in CNT production [43, 54, 55, 58, 61, 62]. These processes have been employed both in single-stage [95–97] and double-stage [55, 59, 61, 74, 98, 99] CNT production in reactors which include: fluidized bed reactors [49, 62, 77], fixed bed reactors [54, 57, 59, 98, 100], amongst several others [43, 64, 95, 99]. Figure 6.4 shows a schematic setup of a double-stage (gasification) production process while Figure 6.5 shows a screw kiln pyrolysis-distillation-catalysis process setup.

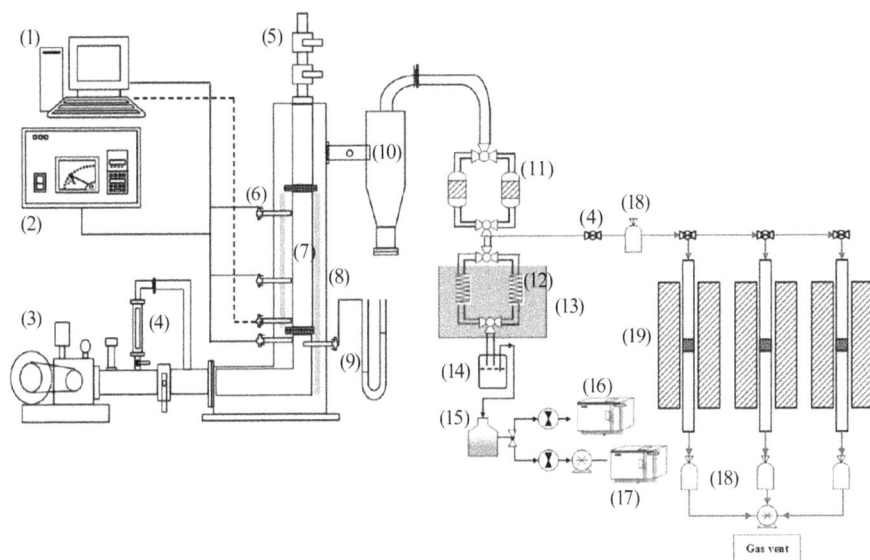

FIGURE 6.4 A schematic setup of CNT production from a fluidized bed gasification process. (1) computer, (2) TIC, (3) blower, (4) flow meter, (5) feeder, (6) thermocouple, (7) sand bed, (8) electric resistance, (9) U-manometer, (10) cyclone, (11) column filter, (12) trapping tube, (13) cooler, (14) gas-washing bottles, (15) backup absorber, (16) GC/TCD, (17) GC/FID, (18) sampling place, and (19) catalysis reactor.

(Reprinted from [100] Copyright (2015) with permission from American Chemical Society).

6.3.2 CARBON NANOTUBE PRODUCTION OPTIMIZATION

The process of choosing the best operating parameters for a process that will optimize the output of the target product and gasification efficiency while lowering overall costs is known as process parameter optimization [83]. The optimization of a full CNT production process in a single stage is quite complex, nonetheless, some authors have carried this out while others have employed a two-stage breakdown for optimum production as well as for coproduction of valuable byproducts. In order to optimize CNT production from solid wastes such as organic and inorganic wastes, several researchers [43, 55–60, 66, 100] have suggested a valuable coproduct such as hydrogen gas. Most of these studies have gone past the laboratory-scale experimentation with different types of catalysts to pilot scale to maximize the production process [70, 102]. Of all these studies, Yang *et al.* [100] optimized CNT production in both production stages (gasification and CVD) stages. In the first stage, waste plastics (a combination of polyethylene and polypropylene) as the carbon source decomposed to CNT and light hydrocarbons in the presence of Ni/Al-SBA-15 catalyst. Meanwhile, in the second stage, the hydrocarbon combines with another catalyst (Ni-Cu/CaO-SiO$_2$) to produce CNT along with hydrogen gas. The CNT fraction obtained in both cases was around 48 wt.% along with very

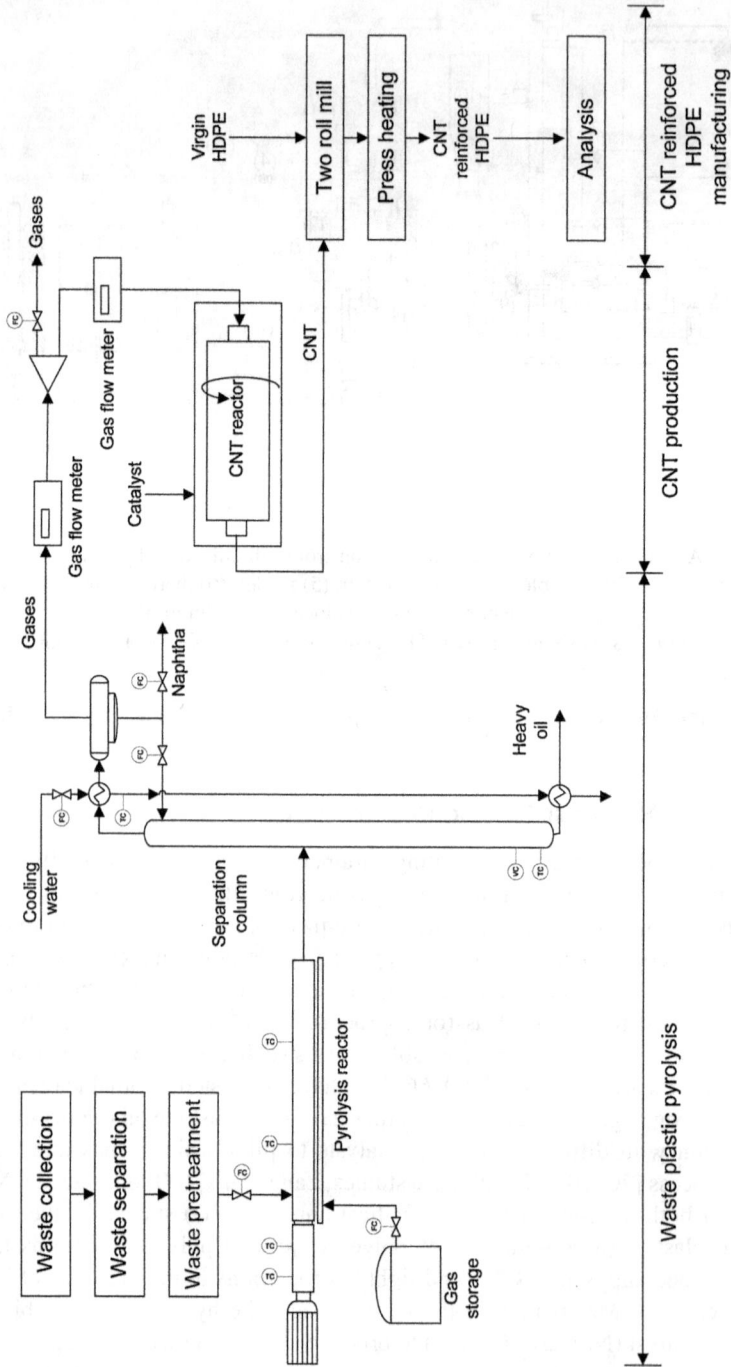

FIGURE 6.5 A setup for CNT production in a screw kiln pyrolysis–distillation–catalysis process. (Reprinted from [101] Copyright (2016) with permission from Elsevier).

high hydrogen with 857.6 mmol/h-g catalyst when the production process was compared and validated with similar studies. The authors concluded that optimum operating conditions such as operating temperature and equivalence ratio all play a crucial role in the process.

Yahyazadeh *et al.* [103] employed Response Surface Methodology (RSM) [104] to optimize CNT growth. In the study, acetylene and Fe/CaCO$_3$ were used as the catalyst and hydrocarbon source respectively in the catalytic CVD method to create CNTs. The CVD reactor's ideal operating conditions for maximizing CNT production and purity were identified using the RSM approach. The authors employed RSM central composite design (CCD) to investigate the effects of the reaction time (30–60 min), temperature (700–800°C), and catalyst loading (10–30 wt.% Fe). From the result, the most notable influences on CNTs growth were determined to be the reaction temperature and reaction time. The highest CNT yields were obtained by using 10Fe/CaCO$_3$ at the minimum reaction temperature (700°C) and the shortest reaction time (both of which were 30 minutes). However, according to Raman spectroscopy data, the CNT quality is not at its best under these circumstances ($I_D/I_G = 0.816$). The authors further advised using a reaction temperature of 800°C to produce high-quality CNTs with a low I_D/I_G ratio, as this is related to the creation of graphitic carbon structure rather than amorphous carbon at high temperatures.

In addition, a number of optimization techniques, including Newton-Raphson [105], Gauss-Newton, numerical methods [106], and Lagrange multipliers are accessible for adoption in CNT production optimization. To determine the ideal process parameter combination, statistical optimization techniques such as Taguchi [107, 108], and Factorial [109–113], Design of Experiment (DOE) [114–117] could also be applied.

6.4   CONCLUSION

Nanomaterials (NMTs) have been around for centuries; however, years of intensive research has provided a better understanding of the mechanisms involved in modeling the production processes involved in their production. Extensive investigations are still needed on the production of aligned NMTs (mostly importantly, CNT) and the regulation of their growth. As a result, there is a vast body of theoretical or experimental research devoted to understanding the nucleation and growth of NMTs. The modeling and optimization of the NMT production process is an interdisciplinary field combining physics, chemistry, biology, material science as well as engineering. In order to design products that are directly applicable to the industrial sector and to stay informed about current demands and upcoming difficulties, industry and academia must work closely together to make NMT production successful. Choosing the right model that combines the NMT production with the necessary characteristics and constrained impurities as well as the approach's scalability is of the utmost importance. In the industrial sphere, the utilization of high-quality NMTs is rapidly expanding, making them the next generation's attractive resources with promising applications. Undoubtedly, in the upcoming years, the current gap between basic research on NMTs and their practical use will be closed.

REFERENCES

1. Tsuzuki, T., *Nanotechnology commercialization*. 2013: CRC Press. p. 480.
2. Nanotechnologies, *Vocabulary–Part 1: Core Terms*. International Standardisation Organisation, Technical Specification ISO/TS, 2015: p. 80004.
3. Zhao, J. and F. Djurabekova, Computational Modeling of Nanoparticles in Inert Environment, in *Frontiers of Nanoscience*. 2020, Elsevier. p. 5–26.
4. Tsuzuki, T., Properties of nanoparticulate materials, in *Handbook of Clinical Nanomedicine: Nanoparticles, Imaging, Therapy, Clinical Applications*, Vol. 1. 2016. p. 171.
5. Mekuye, B. and B. Abera, Nanomaterials: An overview of synthesis, classification, characterization, and applications. *Nano Select*, p. 486–501. 2023. DOI: 10.1002/nano.202300038
6. Baig, N., I. Kammakakam, and W.J.M.A. Falath, Nanomaterials: A review of synthesis methods, properties, recent progress, and challenges. *Materials Advances*, 2021. **2**(6): p. 1821–1871.
7. Talebian, S., et al., Nanotechnology-based disinfectants and sensors for SARS-CoV-2. *Nature Nanotechnology*, 2020. **15**(8): p. 618–621.
8. Nanomedicine and the COVID-19 vaccines. *Nature Nanotechnology*, 2020. **15**(12): p. 963–963.
9. Mohamed, N.A., et al. Nanomedicine as a Potential Tool against Monkeypox. *Vaccines*, 2023. **11**. DOI: 10.3390/vaccines11020428
10. Wang, X.-Y., A. Narita, and K. Müllen, Precision synthesis versus bulk-scale fabrication of graphenes. *Nature Reviews Chemistry*, 2017. **2**(1): p. 1–10.
11. Raji, K. and C.B. Sobhan, Simulation and modeling of carbon nanotube synthesis: Current trends and investigations. *Nanotechnology Reviews*, 2013. **2**(1): p. 73–105.
12. Ijaz, I., et al., Detail review on chemical, physical and green synthesis, classification, characterizations and applications of nanoparticles. *Green Chemistry Letters and Reviews*, 2020. **13**(3): p. 223–245.
13. Wang, Z., et al., Processing and property investigation of single-walled carbon nanotube (SWNT) buckypaper/epoxy resin matrix nanocomposites. *Composites Part A: applied science manufacturing Review*, 2004. **35**(10): p. 1225–1232.
14. Mohan, N., et al., A review on composite materials and process parameters optimisation for the fused deposition modelling process. *Virtual Physical Prototyping*, 2017. **12**(1): p. 47–59.
15. Charitidis, C.A., et al., Manufacturing nanomaterials: From research to industry. *Manufacturing Review*, 2014. **1**: p. 11.
16. Mahmoud, A.E.D., M. Fawzy, and N. Khan, *Artificial Intelligence and modeling for Water Sustainability: Global Challenges*. 1st Edition ed. 2023: CRC Press.
17. Gu, Q., et al., High aspect ratio metamaterials and their applications. *Sensors and Actuators A: Physical*, 2022. **335**: p. 113376.
18. Khan, Y., et al. Classification, synthetic, and characterization approaches to nanoparticles, and their applications in various fields of nanotechnology: A review. *Catalysts*, 2022. **12**. DOI: 10.3390/catal12111386
19. Khairiah, et al., The Electrical properties of nanomaterials derived from durian skin waste by using a various types of electrodes for bio-battery application. *Journal of Physics: Conference Series*, 2018. **1120**(1): p. 012085.
20. Mahmoud, A.E.D., Nanomaterials: Green Synthesis for Water Applications, in *Handbook of Nanomaterials and Nanocomposites for Energy and Environmental Applications*, O.V. Kharissova, L.M.T. Martínez, and B.I. Kharisov, Editors. 2020. Springer International Publishing. p. 1–21.

21. Yang, M. and J. He, Tailoring the structure of metal oxide nanostructures towards enhanced sensing properties for environmental applications. *Journal of Colloid Interface Science*, 2012. **368**(1): p. 41–48.

22. Boustani, I., *Molecular Modelling and Synthesis of Nanomaterials: Applications in Carbon-and Boron-based Nanotechnology*. Vol. 290. p. 555–5889. 2020: Springer Nature.

23. Mahmoud, A.E.D., et al., Green copper oxide nanoparticles for lead, nickel, and cadmium removal from contaminated water. *Scientific Reports*, 2021. **11**(1): p. 12547.

24. Bertolini, T., D. Fungaro, and A. Mahmoud, The influence of separately and combined bentonite and kaolinite as binders for pelletization of NaA zeolite from coal fly ash. *Cerâmica*, 2022. **68**: p. 375–384.

25. Mahmoud, A.E.D., Recent Advances of TiO$_2$ Nanocomposites for Photocatalytic Degradation of Water Contaminants and Rechargeable Sodium Ion Batteries, in *Advances in Nanocomposite Materials for Environmental and Energy Harvesting Applications*, A.E. Shalan, A.S. Hamdy Makhlouf, and S. Lanceros-Méndez, Editors. 2022, Springer International Publishing. p. 757–770.

26. El-Kady, M.M., et al., Nanomaterials: A comprehensive review of applications, toxicity, impact, and fate to environment. *Journal of Molecular Liquids*, 2023. **370**: p. 121046.

27. Huang, L., et al., Enhanced water purification via redox interfaces created by an atomic layer deposition strategy. *Environmental Science: Nano*, 2021. **8**(4): p. 950–959.

28. Rogach, A.L., *Semiconductor nanocrystal quantum dots*. Verlag. 2008: Springer Vienna. 372.

29. Hu, Y. and C. Guo, *Carbon nanotubes and carbon nanotubes/metal oxide heterostructures: synthesis, characterization and electrochemical property*. In: Carbon Nanotubes-Growth: Applications in Energy Combustion Science, 2011: InTech: Rijeka, Croatia, 2011; p. 3–34. DOI: 10.5772/16463

30. Peng, X., et al., Carbon nanotube–nanocrystal heterostructures. *Chemical Society Reviews*, 2009. **38**(4): p. 1076–1098.

31. Madani, S.Y., et al., A new era of cancer treatment: carbon nanotubes as drug delivery tools. *International Journal of Nanomedicine*, 2011: p. 2963–2979. https://doi.org/10.2147/IJN.S16923

32. Mammeri, F., et al., Photoluminescent properties of new quantum dot nanoparticles/carbon nanotubes hybrid structures. *Colloids Surfaces A: Physicochemical Engineering Aspects*, 2013. **439**: p. 138–144.

33. Liang, A., et al., Recent developments concerning the dispersion methods and mechanisms of graphene. *Coatings*, 2018. **8**(1): p. 33.

34. Sun, T.-Y., et al., Unraveling the strong coupling between graphene/nickel interface and atmospheric adsorbates for versatile realistic applications. *Carbon Trends*, 2021. **2**: p. 100013.

35. Radhi, A., et al., Mechanism and factors influence of graphene-based nanomaterials antimicrobial activities and application in dentistry. *Journal of Materials Research Technology*, 2021. **11**: p. 1290–1307.

36. Mahmoud, A.E.D., A. Stolle, and M. Stelter, Sustainable synthesis of high-surface-area graphite oxide via dry ball milling. *ACS Sustainable Chemistry & Engineering*, 2018. **6**(5): p. 6358–6369.

37. Mahmoud, A.E.D., Graphene-based nanomaterials for the removal of organic pollutants: Insights into linear versus nonlinear mathematical models. *Journal of Environmental Management*, 2020. **270**: p. 110911.

38. Mahmoud, A.E.D., et al., Mechanochemical versus chemical routes for graphitic precursors and their performance in micropollutants removal in water. *Powder Technology*, 2020. **366**: p. 629–640.

39. Guadagno, L., et al., The role of carbon nanofiber defects on the electrical and mechanical properties of CNF-based resins. *Nanotechnology*, 2013. **24**(30): p. 305704.

40. Phiri, J., P. Gane, and T.C. Maloney, General overview of graphene: Production, properties and application in polymer composites. *Materials Science Engineering: B*, 2017. **215**: p. 9–28.

41. Generalic, E. *Fullerene*. 2022 Nov 20, 2022; June 29, 2022:[Available from: https://glossary.periodni.com/glossary.php?en=fullerene

42. Zhao, Y.-L. and J.F. Stoddart, Noncovalent functionalization of single-walled carbon nanotubes. *Accounts of Chemical Research*, 2009. **42**(8): p. 1161–1171.

43. Williams, P.T., Hydrogen and carbon nanotubes from pyrolysis-catalysis of waste plastics: A review. *Waste Biomass Valorization*, 2021. **12**(1): p. 1–28.

44. Esfanjani, M. and I.S. Guyo, Modeling of carbon nanotubes (CNTs) and usage in drug delivery. *Technium BioChemMed*, 2022. **3**(3): p. 66–74.

45. Bassyouni, M., et al., Utilization of carbon nanotubes in removal of heavy metals from wastewater: A review of the CNTs' potential and current challenges. *Applied Physics A*, 2020. **126**(1): p. 1–33.

46. Kumar, A., K. Sharma, and A.R. Dixit, Carbon nanotube-and graphene-reinforced multiphase polymeric composites: review on their properties and applications. *Journal of Materials Science*, 2020. **55**(7): p. 2682–2724.

47. Yang, Z., et al., Carbon nanotube-and graphene-based nanomaterials and applications in high-voltage supercapacitor: A review. *Carbon*, 2019. **141**: p. 467–480.

48. Saravanan, M., et al., Techno-economics of carbon nanotubes produced by open air arc discharge method. *International Journal of Engineering, Science Technology*, 2010. **2**(5): p. 100–108.

49. Danafar, F., et al., Mathematical modeling of carbon nanotubes formation in fluidized bed chemical vapor deposition. *International Journal of Chemical Reactor Engineering*, 2017. **15**(2).

50. Su, X., et al., A comparative study of polymer nanocomposites containing multi-walled carbon nanotubes and graphene nanoplatelets. *Nano Materials Science*, 2022. **4**(3): p. 185–204.

51. Awan, U., *A Review of Multi-Walled Carbon Nanotube-Based Interconnects in VLSI Systems: Advantages, Challenges, and Future Directions*. 2023.

52. Nag, A., et al., Multi-walled carbon nanotubes-based sensors for strain sensing applications. *%J Sensors*, 2021. **21**(4): p. 1261.

53. Moothi, K., et al., Coal as a carbon source for carbon nanotube synthesis. *Carbon*, 2012. **50**(8): p. 2679–2690.

54. Wu, C., et al., *Production and application of carbon nanotubes, as a co-product of hydrogen from the pyrolysis-catalytic reforming of waste plastic. Process Safety and Environmental Protection*, 2016. **103**: p. 107–114.

55. Yang, R.-X., et al., Co-production of carbon nanotubes and hydrogen from waste plastic gasification in a two-stage fluidized catalytic bed. *Renewable Energy*, 2020. **159**: p. 10–22.

56. Acomb, J.C., C. Wu, and P.T. Williams, Control of steam input to the pyrolysis-gasification of waste plastics for improved production of hydrogen or carbon nanotubes. *Applied Catalysis B: Environmental*, 2014. **147**: p. 571–584.

57. Zhang, Y. and P.T. Williams, Carbon nanotubes and hydrogen production from the pyrolysis catalysis or catalytic-steam reforming of waste tyres. *Journal of Analytical and Applied Pyrolysis*, 2016. **122**: p. 490–501.

58. Li, W., et al., Catalysts evaluation for production of hydrogen gas and carbon nanotubes from the pyrolysis-catalysis of waste tyres. *International Journal of Hydrogen Energy*, 2019. **44**(36): p. 19563–19572.

59. Zhang, Y., et al., Pyrolysis–catalytic reforming/gasification of waste tires for production of carbon nanotubes and hydrogen. *Energy & Fuels*, 2015. **29**(5): p. 3328–3334.

60. Awadallah, A., et al., Effect of structural promoters on the catalytic performance of cobalt-based catalysts during natural gas decomposition to hydrogen and carbon nanotubes. *Fullerenes, Nanotubes and Carbon Nanostructures*, 2016. **24**(3): p. 181–189.

61. Aboul-Enein, A.A., et al., Synthesis of multi-walled carbon nanotubes via pyrolysis of plastic waste using a two-stage process. *Fullerenes, Nanotubes and Carbon Nanostructures*, 2018. **26**(7): p. 443–450.

62. Zhang, B., et al., Synthesis of carbon nanotubes from conventional biomass-based gasification gas. *Fuel Processing Technology*, 2018. **180**: p. 105–113.

63. Simate, G.S., et al., Kinetic model of carbon nanotube production from carbon dioxide in a floating catalytic chemical vapour deposition reactor. *RSC Advances*, 2014. **4**(19): p. 9564–9572.

64. Samandari-Masouleh, L., et al., Modeling the growth of carbon nanotubes in a floating catalyst reactor. *Industrial Engineering Chemistry Research*, 2012. **51**(3): p. 1143–1149.

65. Das, N., et al., The effect of feedstock and process conditions on the synthesis of high purity CNTs from aromatic hydrocarbons. *Carbon*, 2006. **44**(11): p. 2236–2245.

66. Wu, C., J. Huang, and P.T. Williams, Carbon nanotubes and hydrogen production from the reforming of toluene. *International Journal of Hydrogen Energy*, 2013. **38**(21): p. 8790–8797.

67. Khedr, M., K.A. Halim, and N. Soliman, Effect of temperature on the kinetics of acetylene decomposition over reduced iron oxide catalyst for the production of carbon nanotubes. *Applied Surface Science*, 2008. **255**(5): p. 2375–2381.

68. Stadermann, M., et al., Mechanism and kinetics of growth termination in controlled chemical vapor deposition growth of multiwall carbon nanotube arrays. *Nano Letters*, 2009. **9**(2): p. 738–744.

69. Wirth, C.T., et al., Diffusion-and reaction-limited growth of carbon nanotube forests. *ACS Nano*, 2009. **3**(11): p. 3560–3566.

70. Liu, W., et al., Co-production of hydrogen and carbon nanotubes from cracking of waste cooking oil model compound over Pt–Ni/SBA-15 catalysts. *Journal of Porous Materials*, 2022. **29**(1): p. 49–61.

71. Sha, X. and L. Zhao, Mathematical model and simulation calculation method based on the exfoliation of single-layer graphene from dispersed carbon nanotubes. *Journal of Nanomaterials*, 2022. **2022**: p. 1–11. https://doi.org/10.1155/2022/8503507

72. Xu, X., B. Karami, and D. Shahsavari, Time-dependent behavior of porous curved nanobeam. *International Journal of Engineering Science*, 2021. **160**: p. 103455.

73. Kavinkumar, T. and S. Manivannan, Uniform decoration of silver nanoparticle on exfoliated graphene oxide sheets and its ammonia gas detection. *Ceramics International*, 2016. **42**(1): p. 1769–1776.

74. Radocea, A., et al., Solution-synthesized chevron graphene nanoribbons exfoliated onto H: Si (100). *Nano Letters*, 2017. **17**(1): p. 170–178.

75. Das, R., Nanohybrid Catalyst based on Carbon Nanotube, in *Carbon Nanostructures*. 2017, Springer. p. 23–54.

76. Dasgupta, K., J.B. Joshi, and S. Banerjee, Fluidized bed synthesis of carbon nanotubes–A review. *Chemical Engineering Journal*, 2011. **171**(3): p. 841–869.

77. Philippe, R., et al., Kinetic modeling study of carbon nanotubes synthesis by fluidized bed chemical vapor deposition. *AIChE journal*, 2009. **55**(2): p. 465–474.

78. Raji, K., S. Thomas, and C. Sobhan, A chemical kinetic model for chemical vapor deposition of carbon nanotubes. *Applied Surface Science*, 2011. **257**(24): p. 10562–10570.

79. Kwok, C.T., et al., Temperature and time dependence study of single-walled carbon nanotube growth by catalytic chemical vapor deposition. *Carbon*, 2010. **48**(4): p. 1279–1288.

80. Mahmoud, A.E.D., N.A. Khan, and Y.-T. Hung, *4 Advances in Artificial Intelligence Applications in Sustainable Water Remediation.* Artificial Intelligence and Modeling for Water Sustainability: Global Challenges, 2023: p. 53–69.

81. Onojake, M.C., C. Obi, and A.E.D. Mahmoud, Water Quality Monitoring Using Sensors and Models, in *Artificial Intelligence and Modeling for Water Sustainability.* 2023, CRC Press. p. 97–127.

82. Mahmoud, A.E.D. and P. Krasucka, Global Water Challenges and Sustainability, in *Artificial Intelligence and Modeling for Water Sustainability.* 2023, CRC Press. p. 1–12.

83. Okolie, J.A., et al., Modeling and process optimization of hydrothermal gasification for hydrogen production: A comprehensive review. *The Journal of Supercritical Fluids,* 2021. **173**: p. 105199.

84. Nasr, M., et al., Artificial intelligence modeling of cadmium(II) biosorption using rice straw. *Applied Water Science,* 2017. **7**(2): p. 823–831.

85. Vivanco-Benavides, L.E., et al., Machine learning and materials informatics approaches in the analysis of physical properties of carbon nanotubes: A review. *Computational Materials Science,* 2022. **201**: p. 110939.

86. Kotzabasaki, M., et al., Machine learning methods for multi-walled carbon nanotubes (MWCNT) genotoxicity prediction. *Nanoscale Advances,* 2021. **3**(11): p. 3167–3176.

87. Ho, N.X., T.-T. Le, and M.V. Le, Development of artificial intelligence based model for the prediction of Young's modulus of polymer/carbon-nanotubes composites. *Mechanics of Advanced Materials Structures,* 2022. **29**(27): p. 5965–5978.

88. Le, T.-T., Prediction of tensile strength of polymer carbon nanotube composites using practical machine learning method. *Journal of Composite Materials,* 2020. **55**(6): p. 787–811.

89. Westing, N.M., Carbon Nanotube Growth Rate Regression using Support Vector Machines and Artificial Neural Networks, in *Department of Electrical and Computer Engineering, Graduate School of Engineering and Management.* 2014, Air Force Institute of Technology. p. 131.

90. Hajilounezhad, T., et al. Exploration of carbon nanotube forest synthesis-structure relationships using physics-based simulation and machine learning. in *2019 IEEE Applied Imagery Pattern Recognition Workshop (AIPR).* 2019. IEEE.

91. Bian, L., et al., Machine-learning identification of the sensing descriptors relevant in molecular interactions with metal nanoparticle-decorated nanotube field-effect transistors. *ACS Applied Materials Interfaces,* 2018. **11**(1): p. 1219–1227.

92. Rahmat, M. and P. Hubert, Carbon nanotube–polymer interactions in nanocomposites: A review. *Composites Science Technology,* 2011. **72**(1): p. 72–84.

93. Rahmat, M. and P. Hubert, Molecular dynamics simulation of single-walled carbon nanotube–PMMA interaction *Journal of Nano Research,* 2012. **18**: p. 117–128.

94. Rahmat, M., P. Hubert, Carbon nanotube–polymer interactions in nanocomposites: A review *Composites Science Technology,* 2011. **72**(1): p. 72–84. DOI: 10.1016/j. compscitech.2011.10.002

95. Mishra, N., et al., Pyrolysis of waste polypropylene for the synthesis of carbon nanotubes. *Journal of Analytical Applied Pyrolysis,* 2012. **94**: p. 91–98.

96. Gong, J., et al., Catalytic conversion of linear low density polyethylene into carbon nanomaterials under the combined catalysis of Ni_2O_3 and poly (vinyl chloride). *Chemical engineering journal,* 2013. **215**: p. 339–347.

97. Gong, J., et al., Catalytic carbonization of polypropylene by the combined catalysis of activated carbon with Ni_2O_3 into carbon nanotubes and its mechanism. *Applied Catalysis A: General,* 2012. **449**: p. 112–120.

98. Veksha, A., A. Giannis, and V.W.-C. Chang, Conversion of non-condensable pyrolysis gases from plastics into carbon nanomaterials: Effects of feedstock and temperature. *Journal of Analytical Applied Pyrolysis*, 2017. **124**: p. 16–24.

99. Aboul-Enein, A.A., et al., Simple method for synthesis of carbon nanotubes over Ni-Mo/Al$_2$O$_3$ catalyst via pyrolysis of polyethylene waste using a two-stage process. *Fullerenes, Nanotubes, and Carbon Nanostructures*, 2017. **25**(4): p. 211–222.

100. Yang, R.-X., K.-H. Chuang, and M.-Y. Wey, Effects of nickel species on Ni/Al$_2$O$_3$ catalysts in carbon nanotube and hydrogen production by waste plastic gasification: bench- and pilot-scale tests. *Energy Fuels*, 2015. **29**(12): p. 8178–8187.

101. Borsodi, N., et al., Carbon nanotubes synthetized from gaseous products of waste polymer pyrolysis and their application. *Journal of Analytical Applied Pyrolysis*, 2016. **120**: p. 304–313.

102. Qi, T., et al., Catalytic synthesis of carbon nanotubes by Ni/ZSM-5 catalyst from waste plastic syngas. *Journal of Biobased Materials Bioenergy*, 2022. **16**(3): p. 356–366.

103. Yahyazadeh, A., et al., Optimization of carbon nanotube growth via response surface methodology for Fischer-Tropsch synthesis over Fe/CNT catalyst. *Catalysis Today*, 2022. **404**: p. 117–131.

104. Batool, M., et al., Response surface methodology modeling correlation of polymer composite carbon nanotubes/chitosan nanofiltration membranes for water desalination. *ACS ES&T Water*, 2023. **3**(5): p. 1406–1421.

105. Leonetti, L., et al., Optimal design of CNT-nanocomposite nonlinear shells. *Nanomaterials*, 2020. **10**(12): p. 2484.

106. Cho, J. and H. Kim, Numerical optimization of CNT distribution in functionally graded CNT-reinforced composite beams. *Polymers*, 2022. **14**(20): p. 4418.

107. Qistina, O., et al., Optimization of carbon nanotube-coated monolith by direct liquid injection chemical vapor deposition based on taguchi method. Catalysts, 2020. **10**(1): p. 67.

108. Santangelo, S., et al., Optimization of CVD growth of CNT-based hybrids using the Taguchi method. *Materials Research Bulletin*, 2012. **47**(3): p. 595–601.

109. Zhao, W., et al. Optimization of adsorption conditions using response surface methodology for Tetracycline removal by MnFe$_2$O$_4$/multi-wall carbon nanotubes. *Water*, 2023. **15**. DOI: 10.3390/w15132392

110. Mohammed, I.H.A., et al., Full factorial design approach to carbon nanotubes synthesis by CVD method in argon environment. *South African Journal of Chemical Engineering*, 2017. **24**: p. 17–42.

111. Mubarak, N., et al., Optimization of CNTs production using full factorial design and its advanced application in protein purification. *International Journal of Nanoscience*, 2010. **9**(3): p. 181–192.

112. Mahmoud, A.E.D., M. Fawzy, and A. Radwan, Optimization of Cadmium (CD2+) removal from aqueous solutions by novel biosorbent. *International Journal of Phytoremediation*, 2016. **18**(6): p. 619–625.

113. Badr, N.B.E., K.M. Al-Qahtani, and A.E.D. Mahmoud, Factorial experimental design for optimizing selenium sorption on Cyperus laevigatus biomass and green-synthesized nano-silver. *Alexandria Engineering Journal*, 2020. **59**(6): p. 5219–5229.

114. Chaudhari, R., et al., Experimental investigations and optimization of MWCNTs-mixed WEDM process parameters of nitinol shape memory alloy. *Journal of Materials Research and Technology*, 2021. **15**: p. 2152–2169.

115. Mahmoud, A.E.D., et al., Biogenic synthesis of reduced graphene oxide from Ziziphus spina-christi (Christ's thorn jujube) extracts for catalytic, antimicrobial, and antioxidant potentialities. *Environmental Science and Pollution Research*, 2022. 29: 89772–89787.

116. Mahmoud, A.E.D., M. Franke, and P. Braeutigam, Experimental and modeling of fixed-bed column study for phenolic compounds removal by graphite oxide. *Journal of Water Process Engineering*, 2022. **49**: p. 103085.

117. Mahmoud, A.E.D., et al., Facile synthesis of reduced graphene oxide by Tecoma stans extracts for efficient removal of Ni (II) from water: batch experiments and response surface methodology. *Sustainable Environment Research*, 2022. **32**(1): p. 22.

7 Machine Learning Applications for Nano-synthesized Materials Production and Utilization

Zainab T. Yaqub and Bilainu O. Oboirien

7.1 INTRODUCTION

Although there is no internationally accepted definition of a nanomaterial, a material is considered nano if at least one of its dimensions falls within the range of 1–100 nm. By altering the form and size at the nanoscale level, nanomaterials take on distinctive characteristics that differ from their bulk material counterparts, boasting new features and capabilities that make them of significant scientific interest. Because of the changes in size and morphology of the nanomaterials compared to the bulk materials, several unique properties are observed. These properties differ from those of the bulk materials and include aspects such as surface area, thermal effects, thermal and electrical conductivity, mechanical qualities, catalyst support, antimicrobial activities, and magnetism [1]. Because of these unique properties, nanomaterials provide significant advantages in cost, performance, and activity over traditional materials. They are widely employed in various fields including catalysis, agricultural and food production, hydrogen production and storage, construction, and more [2, 3]. These materials can be distinctly classified according to their size, characteristics, and constituents. The classifications include carbon-based nanomaterials, metal nanoparticles, nanoporous materials, ultrathin 2-dimensional nanomaterials, and core-shell nanoparticles [4]. Table 7.1 describes the different types of nanomaterials and their applications in different industries.

Nanomaterials can be synthesized through two main approaches. The first involves using crushing, milling, and grinding procedures to break down bulk materials into smaller particles to create nanostructured materials [16]. The second utilizes methods such as self-assembly or coprecipitation to assemble tiny particles into nanomaterials. These methods include techniques like milling, electrospinning, chemical vapor deposition, hydrothermal synthesis, and green synthesis. Among all the methods of nanomaterial synthesis, green synthesis is considered to be the most environmentally friendly. It offers many potential applications in biomedical and environmental fields and can also eliminate many hazardous properties by enabling

DOI: 10.1201/9781003371007-7

TABLE 7.1

Types of nanomaterials and their applications in different industries

Nanomaterial Types	Examples	Applications	References
Carbon-based nanomaterials	Fullerenes	Drug delivery tool, transfection agent, and theranostic nanomaterial	[5]
	Carbon nanotubes (CNT)	Artificial implants, genetic engineering, and drug delivery	[6]
	Graphene	Targeted drug delivery, energy storage devices, and electrochemical sensors	[7]
	Nanodiamonds	Bioimaging, drug delivery, and biosensing	[8]
	Quantum dots	LED display, energy conversion, and bioimaging	[9]
	Nanocones	Suitable replacement for CNT	[10]
Nanoporous materials	Activated carbon, porous silicas and zeolins	Energy storage and conversion in fuel cells, solar cells, hydrogen storage and supercapacitors, catalysis, and gas purification	[11]
2-dimensional ultrathin nanomaterials	Sillicene	Quantum sensing and energy devices	[12]
	Mxenex	Energy storage, energy conversion, sensing, optoelectronic, and optical devices	[13]
Metal nanostructured material	Submicron-scale pure metals or their compounds	Catalysis, sensors, optics, and protective coatings	[14]
Core-shell nanoparticles	Ag@Si, Cu@Si, Cu@Ni etc.	Biomedicine, electrical and semiconducting materials, and catalysts	[15]

nanoparticles to be synthesized at lower temperatures, pH, and pressure, all at a significantly reduced cost [17]. The different methods of nanomaterial synthesis are presented in Figure 7.1.

Even with the advancements in nanomaterial synthesis and their growing applications across various industries, there remain ongoing concerns regarding the prediction of nanomaterials' physicochemical properties and the optimization of their production methods. Data-driven machine learning approaches have been introduced to tackle these issues. This chapter provides an overview of several machine learning (ML) models – such as artificial neural networks (ANN), support vector machines (SVM), decision trees, convolutional neural networks (CNN), and deep neural networks (DNN) – that have been used successfully to predict different nanomaterial properties based on experimental data. A detailed examination of these models is included, outlining how they are employed to forecast the physicochemical attributes of nanomaterials. The chapter also discusses the current challenges in utilizing these ML tools and explores the necessary changes for machine learning to have a substantial impact on the field of nanomaterial synthesis in the near future.

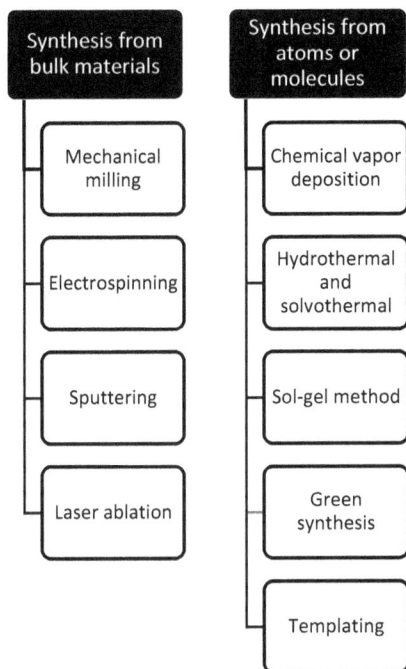

FIGURE 7.1   Different nanomaterial synthesis methods.

(Adapted from [1, 16]).

7.2   MACHINE LEARNING IN NANOMATERIALS

Machine learning (ML) has been applied across various disciplines, making significant contributions to their development. In the realm of nanotechnology, ML has been instrumental in achieving success in predicting properties, optimizing nanomaterial synthesis, and discovering new nanomaterials. The advancements in artificial intelligence have enabled the synthesis and analysis of materials at a pace far exceeding that of traditional one-at-a-time tests, propelling the growth in nanotechnologies. Various ML approaches, including ANN, SVM, decision trees, Gaussian process regression (GPR), Bayesian optimization, convolutional neural networks (CNN), and generative adversarial networks (GAN), have been applied specifically to nanomaterials for property prediction and material discovery [18, 19].

7.2.1   TRADITIONAL ML METHOD IN NANOMATERIALS SYNTHESIS

Traditional ML approaches, which include linear and non-linear regression, ANNs, and decision trees, have been utilized to develop most models of the biological characteristics of nanomaterials [20]. In this section, these approaches will be discussed at length, beginning with an exploration of both linear and non-linear regression techniques.

7.2.1.1　Linear and Non-linear Regression

Linear regression is one of the simplest, yet most powerful statistical methods used to model the relationship between a dependent variable and one or more independent variables. In the context of nanomaterials, it may be applied to predict a biological characteristic based on various physicochemical properties, assuming a linear relationship between them [20]. Unlike linear regression, non-linear regression does not assume a linear relationship between the dependent and independent variables. It can model more complex relationships, making it suitable for the intricate and multifaceted biological characteristics of nanomaterials. This adaptability allows for more accurate modeling of phenomena that are not strictly linear.

7.2.1.2　Artificial Neural Networks (ANNs)

Artificial neural networks (ANNs) are non-linear data processing methods used in various study fields. The neural network can assess different forms of data, discover patterns linked to input-output relationships in the data, and then predict outputs even for new inputs. The ANN model comprises an input layer, a hidden layer, and an output layer. Each neuron layer is made up of a group of neurons. To make reliable predictions, the number of neurons may be arbitrarily adjusted, and prediction model losses can be avoided by altering the number of neurons and activation function parameters. [21]. The ANN modifies the weights and bias parameters to minimize the mean square error (MSE) of the output function. The MSE is defined by Equation (7.1) [22].

$$\text{MSE} = \sum_{i=1}^{n} \frac{\left(x_i - y_i\right)^2}{n} \tag{7.1}$$

where xi is the experimental value, yi is the ANN model's projected value, and n is the number of samples in the dataset.

Among the different applications where ANNs have been deployed, one example is by Orimito et al. [23], who used the ANN tool to determine the relationship between the 3404 input and output data obtained from microreactor-based combinatorial cadmium selenide (CdSe) nanoparticle (NP) synthesis. The input parameters, consisting of reaction temperature, time, amine concentration, Cd concentration, Cd/Se ratio, and type of amine, showed a good relationship with the output parameters, such as particle size, CdSe quantum yield, and photoluminescence quantum yield. The trained data successfully predicted the nanomaterial properties even under new experimental conditions.

This machine learning tool has also been used to predict several other nanomaterial synthesis parameters, such as the diameter of a metallic NP (AgNP), which was produced by green synthesis using cashew gum solution. The variable input conditions such as temperature, time, pH, and concentration of the cashew gum solution were examined [24]. Similarly, ANN was used to predict the role of different synthesis parameters, including the concentration of precursors, pH, and operating temperature, on the shape and size of TiO_2 NP [25]. In essence, various nanomaterials from different synthesis methods have been analyzed to predict outputs such as production efficiency [26] and thermal conductivity [27], among others, based on different synthesis conditions like temperature, time, and pH. A strong prediction relationship was obtained with a correlation coefficient greater than 0.9.

7.2.1.3   Support Vector Machines (SVMs)

Like ANNs, support vector machines (SVMs) are also utilized for regression, classification, and optimization problems especially in the field of nanomaterials. Classification is effectively performed using the kernel approach, which implicitly translates a sample's original feature set into a higher-dimensional feature space. A new sample is categorized based on which side of the margin it falls on [28]. The objective of the SVM model algorithm is to ensure that the projected regression model, given by ($f(x) = wx + b$), falls within the required accuracy of the experimental output [29].

The SVM tool has been utilized in various nanomaterial syntheses, such as determining the dye removal ability of nanostructured adsorbent (MnO_2) [30], predicting the toxic levels of several nanomaterials [31], and assessing the crystallization level of metal-organic nano capsules [32]. Based on this analysis, the SVM has proven to be a powerful tool for predicting different nanomaterial characteristics.

However, while the SVM is more promising than the ANN in terms of generalization performance on unseen data, its main downside is its "black box" nature. A SVM learns information as a set of numerical parameters, making it challenging to understand what the SVM is actually calculating. The development of new models or the addition of interpretable components to existing models would enhance the interpretability of the SVM in ML [33].

7.2.1.4   Decision Tree

Another ML classification tool used in nanomaterial synthesis is the decision tree. The decision tree is trained to draw links between experimental variables and structural attributes that are difficult or impossible to obtain from a small quantity of simulated training data, thereby removing the need for time-consuming correlated electron microscopy [34]. It also gives an explicit graphical depiction of the decision thresholds for each attribute, which significantly improves the interpretation of NP structure-activity models. The decision tree model works by selecting numerous cut points to split the variables in the dataset. Loss functions are computed for both portions of the divided data, and each cut point is compared to the cut point that yields the least loss result. The regression tea model is constructed based on the optimal cut point. The loss function is defined as (Equation 7.2) [22]:

$$\text{Loss}\left(y, f\left(x\right)\right) = \left(f\left(x\right) - y\right)^2 \qquad (7.2)$$

Several authors have used decision trees to predict the relationship between input variables in the synthesis and characterization of different nanoparticles. The decision tree was used to estimate the size of gold nanorods based on a simple interaction between the resonance energy and linewidth of the localized surface plasmon resonance as determined by scanning electron microscopy imaging (SEM) [34]. Similarly, the photocatalytic and cytotoxicity of several multicomponent nanomaterials were analyzed using the decision tree model by Mikolajczyk et al. [35]. The authors used a decision stump approach, followed by a regression metaclassifier, to improve the decision stump's performance. A single descriptor distinguishes each decision stump. The decision tree's maximum size was set to four, and the size correlates to the number of decision stumps. The contribution of several physicochemical factors (the content of silver vs. gold/palladium/platinum, solubility, the electronic properties of a sample, and photocatalytic activity) on photocatalytic properties and environmental toxicity was obtained based on the developed model and experimental evidence obtained.

A random forest regression, a subset of the decision trees, works on the premise of developing multiple distinct decision trees, each of which may be trained individually given training samples. The approach is built on ensemble learning, which can help prevent specific overfitting issues and provide the model with more generality and accuracy [36].

Using the random forest regression, Mirzaei et al. [37] predicted the antibacterial capacity of several nanoparticles. With over 1000 rows of experimental data obtained from 60 articles, input variables such as type of NP, different nanospecific distributors, bacteria characteristics and exposure conditions were used to predict different anti-bacterial measurements such as bacteria viability, minimum bactericidal concentration and zone of inhibition, amongst others. Compared with other regression analysis (SVM, elastic net regression (ENR), ridge regression (RR) and lasso regression), the random forest had the highest R^2 value and the lowest MSE as seen in Figure 7.2. The random forest also determined which input variable has the most effect on the anti-bacterial capacity, which is the core size from the nano-specific distributor.

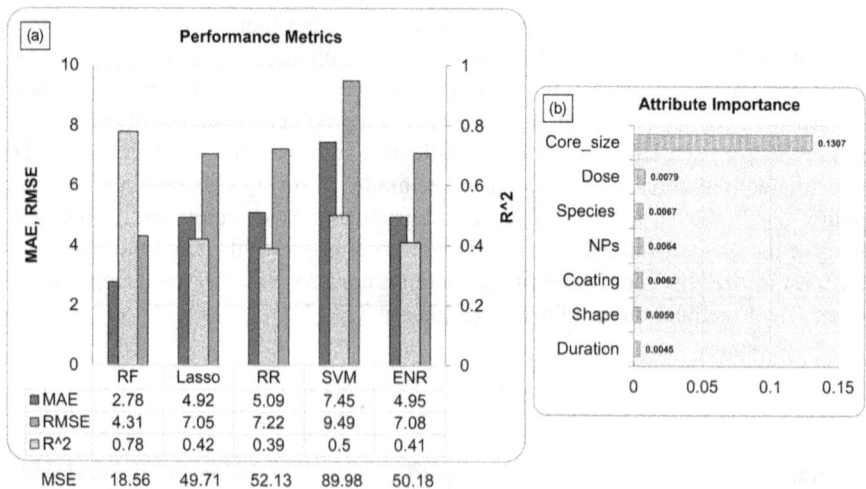

	RF	Lasso	RR	SVM	ENR
■ MAE	2.78	4.92	5.09	7.45	4.95
■ RMSE	4.31	7.05	7.22	9.49	7.08
□ R^2	0.78	0.42	0.39	0.5	0.41
MSE	18.56	49.71	52.13	89.98	50.18

FIGURE 7.2 Performance evaluation of different ML regression analysis models compared with the random forest (RF) [37].

7.2.2 DEEP LEARNING METHODS IN NANOMATERIAL SYNTHESIS

Deep learning approaches use neural networks with more hidden layers and complex architectures with the capacity to identify patterns in images, speech recognition, and dynamic decision making [38]. Deep learning differs from typical machine learning methods in that it employs a hierarchical cascade of non-linear functions. The role of each level is to transform its given data into a conceptual and representation feature that provides input for the succeeding levels. This offers it the comparative benefit of not requiring the scientist to manually conduct feature engineering since the algorithm exhibits self-adjustment efficiency and the ability to identify suitable features independently in a continuous learning process. Convolutional neural networks (CNNs) [39], deep neural networks (DNNs) or autoencoders [40], and generative adversarial networks (GANs) [41] are some of the most extensively used deep learning techniques for nanomaterials synthesis (Figure 7.3).

7.2.2.1 Convolutional Neural Networks (CNN)

A CNN uses several convolutional layers and pooling for complex, multidimensional data. The convolutional layer has a kernel that determines the collected features, and the pooling layer minimizes the number of parameters, lowering the computational complexity even further [19]. Individual neurons which are inside the CNN screens are trained to do mathematical computations and can distinguish tiny, different components of input data that are crucial to a given modeling job. The CNN is mostly

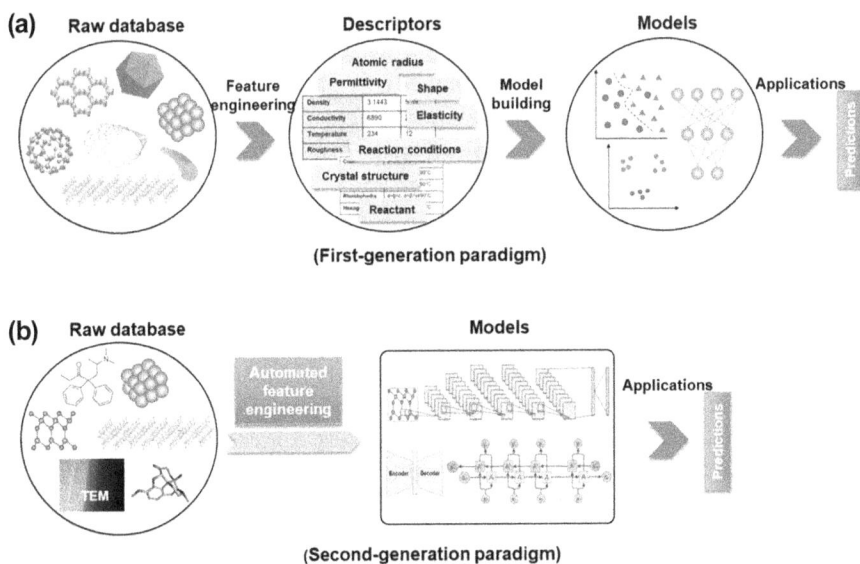

(First-generation paradigm)

(Second-generation paradigm)

FIGURE 7.3 Difference between the traditional ML (a), and deep learning methods (b).

(Copied with permission from Wang et al. [42]).

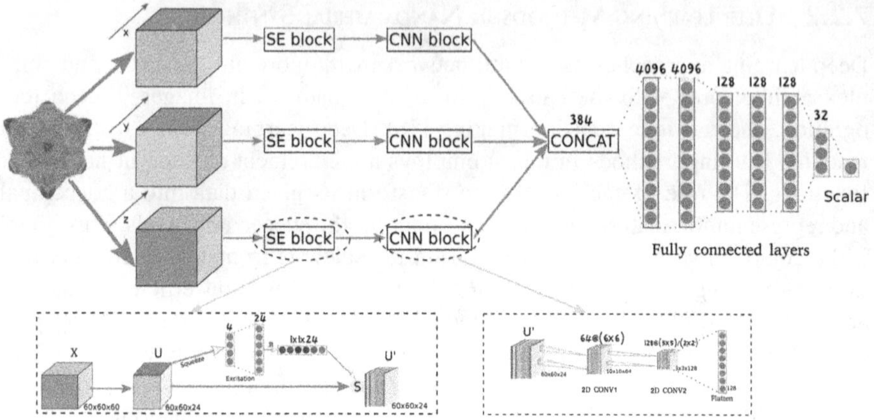

FIGURE 7.4   2D CNN architecture with convolutional, pooling, and fully interconnected layers.

(Reposted with permission from Zhao et al. [45]).

used in nanotechnology to identify nanoparticles in electron microscope pictures (scanning electron microscopy (SEM), signal intensity maps (SIM), transmission electron microscopy (TEM) etc.) [43, 44] (Figure 7.4).

Kolenov et al. [46] used a CNN in classifying and detecting nanoparticles (polystyrene latex nanosphere) deposited in wafers from a SIM. The authors acquired training data from calibrated samples of polystyrene latex (PSL) nanospheres spin-coated on silicon wafers varying in diameter from 40 to 80 nm and investigated the parts of the wafer where the nanoparticle is not present, adding to the "background" class. From the 1302 images given to the network, a dataset with around 260 images per class was constructed, and a 60-20-20 split for training, validation, and testing was implemented. To ensure representative images, the dataset was generated by randomly separating data from each class into three sections and then combining them to generate unbiased sets. The CNN classification method outperformed the traditional ML method for analyzing dispersed maps presented by the authors using modified K-means [47] with an accuracy enhanced by a factor of two for relatively tiny particles with sizes (classes) of 40 and 50 nm.

A CNN was also used for nanostructures in virtual molecular projections [48], spatial distribution maps [49], and TEM images of nanoparticles [43]. This deep learning approach was able to quantitatively analyze a trained CNN with neurons capable of recognizing distinct nanostructure features critical to activities and physicochemical properties, as well as recognizing other nanoscale ordered structures and successfully performing image classification with 94% accuracy. In general, a CNN trained to detect nanoparticles may discover other properties, such as color patches, topography, or backgrounds, that can become highly predictive and be widely employed to design nanoparticles strategically with the desired function.

7.2.2.2 Deep Neural Networks (DNN)

Deep neural networks (DNNs) are composed of two networks: one that maps each material to a result and the other that maps the result to one or more materials. Based on the learned ML model, this deep learning tool, also defined as an autoencoder, may reduce the dimensionality of datasets and forecast materials with specific attributes [41]. A DNN is also helpful for predicting output parameters from input variables when the link between input and output variables is uncertain [50].

The deep learning neural network has been used in several nanomaterials' applications. Amor et al. [51] used the DNN to predict the dye removal ability of nano TiO_2 by mapping out the input-output relationship in order to obtain an optimum result. The input variables consist of the chemicals used in synthesizing the nano TiO_2 and the reaction time. At the same time, the output was the dye-removal efficiency obtained by evaluating the discoloration of methylene blue solution under UV light. An R^2 value greater than 0.99 was obtained with a non-linear relationship between the input and output conditions.

Interestingly, DNN was used to predict the biological fate of gold (Au) NP (spleen and liver accumulation of AuNP) when the nanoparticle was injected into the bloodstream [52]. The dataset included 63,630 proteins' label-free quantitative (LFQ) intensities and five different Au nanoparticle sizes (8–80 nm) which were the input parameters and quantified Au in the blood for over 24 hours & gold content in the liver and spleen that serves as the output. The deep learning model predicted the injected nanoparticle size with over 90% accuracy, liver accumulation with 81%–93% accuracy, spleen uptake with 77%–92.6% accuracy, blood half-life of nanoparticles with 84%–91% accuracy, and biological fate with 77%–94% accuracy.

7.2.2.3 Generative Adversarial Networks (GANs)

Generative Adversarial Networks (GANs) were employed to create nanophotonic structures with exact user-defined spectrum responses. A GAN comprises two parts: a generator that creates trial structure-property models and a discriminator that compares the quality of trial models to existing unlabeled data [53]. Just like DNNs, GANs can also convert structural, physicochemical, and processing parameters in nanomaterials into latent variables/descriptors that may be utilized to create ML models of attributes crucial to the utility and safety of nanomaterials.

Although a GAN has not been individually used in nanomaterial synthesis and application, it has been combined with a CNN to create nanophotonic antennas that are not limited by specified designs [54].

7.3 CHALLENGES AND FUTURE OUTLOOK

Although machine learning has the ability to make predictions with minimum human input and significant time and performance efficiency, it also introduces new hurdles in ensuring forecast accuracy in nanomaterials research. Several challenges have been presented by Brown et al. [41] which include a shortage of datasets that are not constrained to the parameter space of the target issue, fluctuating ML performance owing to the use of simulation data for data training instead of experimental data since it is faster, and difficulties integrating and harmonizing datasets. Apart from the issues of data sets, problems with descriptors are also present for machine learning potentials, particularly for materials with complicated chemical compositions. Because of the necessity for a considerable number of data sets and rising descriptors, training machine learning potentials is exceedingly tough, complicated, time-consuming, and expensive [19].

Furthermore, guaranteeing model performance is crucial because only relevant characteristics are helpful for the creation of prediction models [55]. However, due to the requirements of deep insight into both domain-expert knowledge and the implementation of learning algorithms, selecting the most relevant features is somewhat tricky.

In order to solve the challenge of shortage of datasets, Hiszpanski et al. [56] produced scientific article-processing tools that capture and organize information from nanomaterials publications' text and images, allowing the establishment of a customized knowledge base for nanomaterials synthesis that can be mined to help inform future nanomaterials research. This was accomplished by: (1) creating a relevant corpus of nanomaterials articles, (2) extracting metadata from each article, (3) processing the text of each article to identify the target nanomaterial's morphology and composition, and (4) extracting and further processing SEM/TEM figures of nanomaterials to obtain their morphology and size distributions. This tool has the potential to speed up the discovery of novel nanomaterials. In the future, large nanomaterials databases, such as PubChem or protein data bank (PDB) format, can be used to facilitate rational nanomaterial design and the establishment of data-driven structure-activity relationships, as well as safety, ethics, and regulatory compliance [57].

Also, ML model combinations can be used to swiftly predict the functional parameters of hypothetical NMs at a resolution unattainable by computational or experimental approaches, as well as to discover principles or rules that could govern rational NM and also to detect interior flaws in three-dimensional NMs using computed tomography (CT) images [33].

Finally, future advances in data generation for nanomaterial synthesis via machine learning will guarantee that various scientific and socioeconomic information is combined to detect trends and requirements for society on time and optimize the commercialization benefits of nanotechnology.

7.4 CONCLUSION

Due to the rapid development and synthesis of nanomaterials, faster exploration and characterization of nanomaterials is inevitable. This chapter described several ML tools that have successfully predicted the properties of nanomaterials based on different input properties used in their synthesis. The use of ANNs, SVMs, decision trees, CNNs, and DNNs and the problems associated with their use were all described in detail. Prospective thoughts on changes that must occur in the near future in order to make a more meaningful contribution to nanomaterial synthesis were also presented. Through the aid of ML tools, the development of safe nanomaterials for industrial applications can be expedited with high accuracy.

REFERENCES

[1] N. Baig, I. Kammakakam, W. Falath, and I. Kammakakam, "Nanomaterials: A review of synthesis methods, properties, recent progress, and challenges," *Mater. Adv.*, vol. 2, no. 6, pp. 1821–1871, 2021.

[2] M. Haris et al., "Nanotechnology – A new frontier of nano-farming in agricultural and food production and its development," *Sci. Total Environ.*, vol. 857, 2023.

[3] A. H. Nawar, "Nano-technologies and nanomaterials for civil engineering construction works applications," *Mater. Today Proc.*, 2021.

[4] R. Kumar et al., "Impacts of plastic pollution on ecosystem services, sustainable development goals, and need to focus on circular economy and policy interventions," *Sustain.*, vol. 13, no. 17, 2021.

[5] M. Serda et al., "Synthesis and applications of fullerene nanoconjugate with 5-aminolevulinic acid and its glycoconjugate as drug delivery vehicles," *RSC Adv.*, vol. 12, no. 11, pp. 6377–6388, 2022.

[6] P. R. Kowalski, M. Kasina, and M. Michalik, "Metallic elements fractionation in municipal solid waste incineration residues," *Energy Procedia*, vol. 97, pp. 31–36, 2016.

[7] Mahmoud, A.E.D., "Graphene-based nanomaterials for the removal of organic pollutants: Insights into linear versus nonlinear mathematical models," *J. Environ. Manag.*, 2020. **270**: p. 110911.

[8] J. X. Qin et al., "Nanodiamonds: Synthesis, properties, and applications in nanomedicine," *Mater. Des.*, vol. 210, 2021.

[9] M. A. Cotta, "Quantum dots and their applications: What lies ahead?," *ACS Appl. Nano Mater.*, vol. 3, no. 6, pp. 4920–4924, 2020.

[10] P. Karfa, S. De, K. C. Majhi, R. Madhuri, and P. K. Sharma, "Functionalization of carbon nanostructures," *Compr. Nanosci. Nanotechnol.*, vol. 1–5, pp. 123–144, 2019.

[11] S. Bhattacharyya, Y. Mastai, R. Narayan Panda, S. H. Yeon, and M. Z. Hu, "Advanced nanoporous materials: Synthesis, properties, and applications," *J. Nanomater.*, vol. 2014, 2014.

[12] A. Molle, C. Grazianetti, L. Tao, D. Taneja, M. H. Alam, and D. Akinwande, "Silicene, silicene derivatives, and their device applications," *Chem. Soc. Rev.*, vol. 47, no. 16, pp. 6370–6387, 2018.

[13] L. Jiang et al., "2D single- and few-layered MXenes: synthesis, applications and perspectives," *J. Mater. Chem. A*, vol. 10, no. 26, pp. 13651–13672, 2022.

[14] A. Mitra and G. De, "Sol-Gel synthesis of metal nanoparticle incorporated oxide films on glass," *Glas. Nanocomposites Synth. Prop. Appl.*, pp. 145–163, 2016. https://doi.org/10.1016/B978-0-323-39309-6.00006-7

[15] A. V. Nomoev et al., "Structure and mechanism of the formation of core-shell nanoparticles obtained through a one-step gas-phase synthesis by electron beam evaporation," *Beilstein J. Nanotechnol.*, vol. 6, no. 1, pp. 874–880, 2015.

[16] Mahmoud, A.E.D., A. Stolle, and M. Stelter, "Sustainable synthesis of high-surface-area graphite oxide via dry ball milling," *ACS Sustain. Chem. Eng.*, 2018. **6**(5): p. 6358–6369.

[17] A.E.D. Mahmoud, et al., Green copper oxide nanoparticles for lead, nickel, and cadmium removal from contaminated water. *Sci. Rep.*, 2021. **11**(1): p. 12547.

[18] A.E.D. Mahmoud, M. Fawzy, and N. Khan, *Artificial Intelligence and modeling for Water Sustainability: Global Challenges.* 1st ed. 2023: CRC Press.

[19] W. Liu, Y. Wu, Y. Hong, Z. Zhang, Y. Yue, and J. Zhang, "Applications of machine learning in computational nanotechnology," *Nanotechnology*, vol. 33, no. 16, 2022.

[20] K. G. Reyes and B. Maruyama, "The machine learning revolution in materials?," *MRS Bull.*, vol. 44, no. 7, pp. 530–537, 2019.

[21] R. Liu et al., "Causal inference machine learning leads original experimental discovery in CdSe/CdS core/shell nanoparticles," *J. Phys. Chem. Lett.*, vol. 11, no. 17, pp. 7232–7238, 2020.

[22] X. Chen and H. Lv, "Intelligent control of nanoparticle synthesis on microfluidic chips with machine learning," *NPG Asia Mater.*, vol. 14, no. 1, 2022.

[23] Y. Orimoto et al., "Application of artificial neural networks to rapid data analysis in combinatorial nanoparticle syntheses," *J. Phys. Chem. C*, vol. 116, no. 33, pp. 17885–17896, 2012.

[24] D. B. Calçada, P. V. Quelemes, D. A. Silva, R. A. L. Rabêlo, R. S. A. Leite, and C. Eiras, "An application of artificial neural networks to estimate the diameter of nanoparticles synthesized by a green cashew-gum-based process," *J. Glob. Innov.*, vol. 2, no. 1, pp. 1–11, 2020.

[25] F. Pellegrino et al., "Machine learning approach for elucidating and predicting the role of synthesis parameters on the shape and size of TiO_2 nanoparticles," *Sci. Rep.*, vol. 10, no. 1, 2020.

[26] M. Sadeghzadeh et al., "Prediction of thermo-physical properties of TiO_2-Al_2O_3/water nanoparticles by using artificial neural network," *Nanomaterials*, vol. 10, no. 4, 2020.

[27] Y. Y. Abeska and L. Cavas, "Artificial neural network modelling of green synthesis of silver nanoparticles by honey," *Neural Netw. World*, vol. 32, no. 1, pp. 1–14, 2022.

[28] D. Hejazi, S. Liu, A. Farnoosh, S. Ostadabbas, and S. Kar, "Development of use-specific high-performance cyber-nanomaterial optical detectors by effective choice of machine learning algorithms," *Mach. Learn. Sci. Technol.*, vol. 1, no. 2, 2020.

[29] D. J. Armaghani, P. G. Asteris, B. Askarian, M. Hasanipanah, R. Tarinejad, and V. Van Huynh, "Examining hybrid and single SVM models with different kernels to predict rock brittleness," *Sustain.*, vol. 12, no. 6, pp. 1–17, 2020.

[30] N. M. Mahmoodi, Z. Hosseinabadi-Farahani, and H. Chamani, "Nanostructured adsorbent (MnO_2): Synthesis and least square support vector machine modeling of dye removal," *Desalin. Water Treat.*, vol. 57, no. 45, pp. 21524–21533, 2016.

[31] O. T. Oladele, "Nanomaterials characterization using hybrid genetic algorithm based support vector machines," *Int. J. Mater. Sci. Eng.*, vol. 2, pp. 107–114, 2014.

[32] Y. Xie et al., "Machine learning assisted synthesis of metal-organic nanocapsules," *J. Am. Chem. Soc.*, vol. 142, no. 3, pp. 1475–1481, 2020.

[33] Y. Jia, X. Hou, Z. Wang, and X. Hu, "Machine learning boosts the design and discovery of nanomaterials," *ACS Sustain. Chem. Eng.*, vol. 9, no. 18, pp. 6130–6147, 2021.

[34] K. Shiratori et al., "Machine-learned decision trees for predicting Gold nanorod sizes from spectra," *J. Phys. Chem. C*, vol. 125, no. 35, pp. 19353–19361, 2021.

[35] A. Mikolajczyk, N. Sizochenko, E. Mulkiewicz, A. Malankowska, B. Rasulev, and T. Puzyn, "A chemoinformatics approach for the characterization of hybrid nanomaterials: Safer and efficient design perspective," *Nanoscale*, vol. 11, no. 24, pp. 11808–11818, 2019.

[36] Y. Chen, W. Zheng, W. Li, and Y. Huang, "Large group activity security risk assessment and risk early warning based on random forest algorithm," *Pattern Recognit. Lett.*, vol. 144, pp. 1–5, 2021.

[37] M. Mirzaei, I. Furxhi, F. Murphy, and M. Mullins, "A machine learning tool to predict the antibacterial capacity of nanoparticles," *Nanomaterials*, vol. 11, no. 7, 2021.

[38] D. C. Elton, Z. Boukouvalas, M. D. Fuge, and P. W. Chung, "Deep learning for molecular design - A review of the state of the art," *Mol. Syst. Des. Eng.*, vol. 4, no. 4, pp. 828–849, 2019.

[39] D. Duvenaud, D. Maclaurin, J. A. Iparraguirre, R. Gómez-Bombarelli, T. Hirzel, A. Aspuru-Guzik, and R. P. Adams, "Convolutional Networks on Graphs for Learning Molecular Fingerprints," *J. Chem. Inf. Model.*, vol. 56, no. 2, pp. 399–411, 2016.

[40] S. S. Kalantre et al., "Machine learning techniques for state recognition and auto-tuning in quantum dots," *npj Quantum Inf.*, vol. 5, no. 1, 2019.

[41] K. A. Brown, S. Brittman, N. Maccaferri, D. Jariwala, and U. Celano, "Machine learning in nanoscience: Big Data at small scales," *Nano Lett.*, vol. 20, no. 1, pp. 2–10, 2020.

[42] M. Wang, T. Wang, P. Cai, and X. Chen, "Nanomaterials discovery and design through machine learning," *Small Methods*, vol. 3, no. 5, 2019.

[43] A. Koyama, S. Miyauchi, K. Morooka, H. Hojo, H. Einaga, and Y. Murakami, "Analysis of TEM images of metallic nanoparticles using convolutional neural networks and transfer learning," *J. Magn. Magn. Mater.*, vol. 538, 2021.

[44] A. Cid-Mejías, R. Alonso-Calvo, H. Gavilán, J. Crespo, and V. Maojo, "A deep learning approach using synthetic images for segmenting and estimating 3D orientation of nanoparticles in EM images," *Comput. Methods Programs Biomed.*, vol. 202, 2021.

[45] Y. Zhao, K. Yuan, Y. Liu, S. Y. Louis, M. Hu, and J. Hu, "Predicting elastic properties of materials from electronic charge density using 3D deep convolutional neural networks," *J. Phys. Chem. C*, vol. 124, no. 31, pp. 17262–17273, 2020.

[46] D. Kolenov, D. Davidse, J. Le Cam, and S. F. Pereira, "Convolutional neural network applied for nanoparticle classification using coherent scatterometry data," *Appl. Opt.*, vol. 59, no. 27, p. 8426, 2020.

[47] D. Kolenov and S. F. Pereira, "Machine learning techniques applied for the detection of nanoparticles on surfaces using coherent Fourier scatterometry," *Opt. Express*, vol. 28, no. 13, p. 19163, 2020.

[48] D. P. Russo, X. Yan, S. Shende, H. Huang, B. Yan, and H. Zhu, "Virtual molecular projections and convolutional neural networks for the end-to-end modeling of nanoparticle activities and properties," *Anal. Chem.*, vol. 92, no. 20, pp. 13971–13979, 2020.

[49] Y. Li et al., "Convolutional neural network-assisted recognition of nanoscale L12 ordered structures in face-centred cubic alloys," *npj Comput. Mater.*, vol. 7, no. 1, 2021.

[50] M. H. Ha and O. T. C. Chen, "Deep neural networks using capsule networks and skeleton-based attentions for action recognition," *IEEE Access*, vol. 9, pp. 6164–6178, 2021.

[51] N. Amor, M. T. Noman, and M. Petru, "Prediction of methylene blue removal by nano tio2 using deep neural network," *Polymers (Basel).*, vol. 13, no. 18, 2021.

[52] J. Lazarovits et al., "Supervised learning and mass spectrometry predicts the in vivo fate of nanomaterials," *ACS Nano*, vol. 13, no. 7, pp. 8023–8034, 2019.

[53] Z. Liu, D. Zhu, S. P. Rodrigues, K. T. Lee, and W. Cai, "Generative model for the inverse design of metasurfaces," *Nano Lett.*, vol. 18, no. 10, pp. 6570–6576, 2018.

[54] S. So and J. Rho, "Designing nanophotonic structures using conditional deep convolutional generative adversarial networks," *Nanophotonics*, vol. 8, no. 7, pp. 1255–1261, 2019.

[55] K. T. Butler, D. W. Davies, H. Cartwright, O. Isayev, and A. Walsh, "Machine learning for molecular and materials science," *Nature*, vol. 559, no. 7715, pp. 547–555, 2018.

[56] A. M. Hiszpanski et al., "Nanomaterial synthesis insights from machine learning of scientific articles by extracting, structuring, and visualizing knowledge," *J. Chem. Inf. Model.*, vol. 60, no. 6, pp. 2876–2887, 2020.

[57] Z. Ji, W. Guo, S. Sakkiah, J. Liu, T. A. Patterson, and H. Hong, "Nanomaterial databases: Data sources for promoting design and risk assessment of nanomaterials," *Nanomaterials*, vol. 11, no. 6, 2021.

8 Status and Progress of Nanomaterials Application in Hydrogen Storage

*Fatih Güleç, William Oakley, Xin Liu,
Shahrouz Nayebossadri, Feiran Wang, Emma K.
Smith, Sarah M. Barakat, and Edward H. Lester*

8.1 INTRODUCTION TO HYDROGEN STORAGE IN NANOMATERIALS

Nanostructure materials (with nanoparticles 5–50 nm) offer different chemical, physical, thermodynamic, and hydrogen transport properties compared to their bulk counterparts. Nanostructuring can potentially create new avenues to make a step change in the performance of hydrogen storage materials. Several methods have been developed to synthesize nanostructure materials, offering unique properties which need to be further explored. Nevertheless, to achieve consistent hydrogen absorption and desorption properties, novel methods to stabilize nanoparticles in the highly porous scaffold are still under development. The solid-state hydrogen storage studies have been focused on four main nanomaterials: carbonaceous, metal and complex hydrides, metal-organic frameworks, and covalent organic frameworks.

The first group of nanomaterials are *carbonaceous nanomaterials*, which are considered safe, efficient, and economical candidate solid materials for hydrogen storage due to their unique advantages such as wide availability, low cost, high surface area, thermal stability, and chemical stability. One of the first investigations of the adsorption of hydrogen on carbonaceous materials was reported by Dillon et al. [1]; hydrogen can condense to a high density inside single-walled nanotubes (SWCNTs) with an adsorption capacity of 5–10 wt.% at 133 K and a low hydrogen pressure of 0.04 MPa. Since then, many carbon materials including activated carbons, graphene, and carbon nanotubes have been widely investigated.

The second group of nanomaterials are *metal-organic frameworks (MOFs)*, which are coordination polymers, composed of inorganic metal ions (or clusters) bound to organic linkers. These materials are highly porous with a tunable internal structure, enabling the development of frameworks with a high affinity for storing any targeted species, such as hydrogen gas. This has enabled the storage of large gas volumes at

DOI: 10.1201/9781003371007-8

lower pressures than previously required by traditional compression tanks [2]. The tunability of the MOF materials is derived from a combination of their coordination-based bonding network and the relative ease of manipulating either the organic linker or the metal component. Through chemical functionalization, complete or partial substitution with alternative isostructural building blocks, and careful consideration of complexation and supramolecular assembly chemistry, allows intimate control over the physical, chemical, and structural properties of the resultant framework [3–6]. Practical gas adsorption within MOFs is reliant on the rapid manipulation of operation conditions between two set points, required either for the adsorption or desorption process to occur. These "swing" parameters, normally pressure, temperature or a combination of the two, exploit the additional binding energy sites found within MOF internal pores García-Holley et al. [7]. During adsorption, low temperatures and/or high pressures favor gas retention. Whereas during desorption, the higher temperatures and lower pressures overcome the additional binding energy to favor gas release. However, the challenge for MOF-based hydrogen storage is the limited number of interactions that are exploitable to increase adsorption within the material, due to the homonuclear diatomic nature of H_2 and the consequential absence of any appreciable dipole moment [8].

The next group of nanomaterials for hydrogen storage are *metal and complex hydrides*. Alkali and alkali-earth metals (light metal hydrides) such as Mg, Li, Na, and Ca exhibit highly stable hydride phases along with high desorption temperatures [9]. Strong ionic bonds, high thermodynamic stability, and high energy requirement to break their chemical bonds with hydrogen result in slow H_2 absorption and desorption kinetics, low cyclic stability and reversibility, high activation barriers and low equilibrium pressures [10]. Extensive research in the last decade has led to a rather strong consensus that nanostructure hydrides, specifically obtained by mechanical milling, exhibit substantially better hydrogen storage characteristics. Mechanical milling was proposed to play a dual role by breaking the surface oxide layer and microstructural modification. Nanoparticles could offer significantly higher fresh surface area and grain boundaries, reducing the required activation energy for hydrogenation and therefore significantly improving hydrogen sorption kinetics. Mechanochemical milling i.e. ball milling is the most common method to synthesize metal hydrides hydrogen storage materials Lai et al. [11]. Ball milling is a highly flexible approach for adding dopants, and catalysts and creating large amounts of strain and grain boundaries beneficial to increase hydrogen diffusivity and reaction kinetics [11, 12]. Ball milling of complex metal hydrides has been less effective in reducing particle size compared to binary and intermetallic hydrides Schneemann et al. [13]. However, a range of materials such as amides, alanates, borohydrides, and reactive composite hydrides, have been more recently prepared following the development of more sophisticated milling tools, such as high-pressure vials fitted with pressure and temperature sensors [11].

The final group are *covalent organic frameworks (COFs)* which are constructed from strong covalent bonds of light elements (boron, carbon, nitrogen, oxygen, and silicon) rather than metal ions. The COF attains crystals with a balance between the thermodynamic reversibility of the linking reactions and their kinetics. This class of sorbent materials preserves the advantages of MOFs for hydrogen storage with high surface area, comparable pore dimensions to the length scale of molecular diameter of hydrogen, and rigid structures while the light elements that construct their molecular framework reduce the weight, making COFs excellent candidates for storage of hydrogen.

8.2 CARBONACEOUS NANOMATERIALS

8.2.1 ACTIVATED CARBONS

Activated carbons in a broad sense are made up of a wide range of amorphous carbonaceous materials that possess high porosity and surface area, and which are obtained from the thermal treatment of a variety of carbon materials under different conditions. A variety of raw materials or precursors have been studied, and can be divided into four categories: coal-based materials, biomass, waste materials produced from an industrial process, and synthesized materials. Essentially, the preparation of active carbon normally involves two steps: the carbonization of the raw materials at a certain temperature in an inert atmosphere and the activation of the obtained carbonized materials including physical and chemical activation approaches that could create new pores or remove the blockage in pores and further develop the pore structure Bansal et al. [14]. A schematic of activated carbon preparation is shown in Figure 8.1. Compared to physical activation, chemical activation presents unique advantages such as lower activation temperature, higher surface area and controlled micropore size distribution [15].

Hydrogen adsorption in the porous structure depends on the potential energy resulting from the interaction between a gas molecule and the whole pore structure, which is significantly affected by the pore shape and pore width Lowell et al. [16]. For activated carbon, the adsorption potential of pores originates from the van der Waals potential created by the adjacent pore walls, which is determined by the adsorbate, the adsorption temperature and external pressure. By applying various models, theoretical

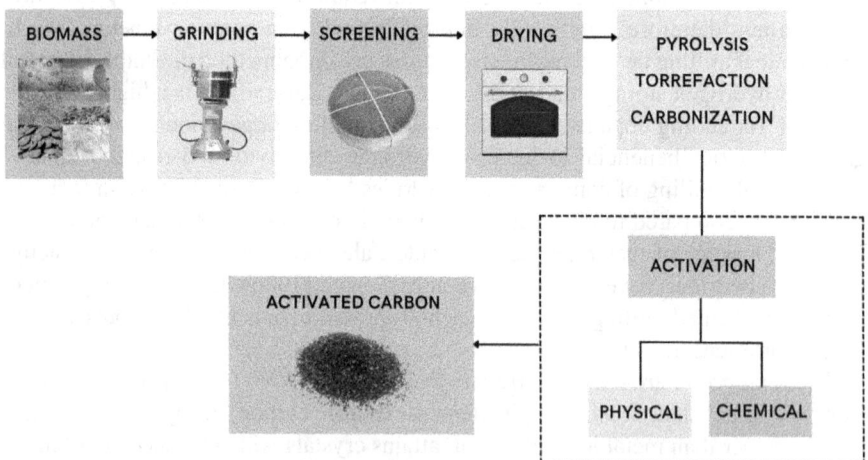

FIGURE 8.1 Schematics of activated carbon synthesis method.

calculations have shown that the optimum width of pores for hydrogen storage at an external pressure of 10 MPa and temperature of 300 K is 0.6–0.7 nm [17, 18]. This is consistent with the experimental results obtained at 293 K and 70 MPa by De la Casa-Lillo et al. [19]. Hydrogen adsorption on most carbons is mainly monolayer, therefore, the maximum theoretical hydrogen storage capacity of carbon materials can be estimated from their surface area (H_2 adsorption capacity = $0.00227*S_{BET}$).

Taking all those parameters into account, activated carbons with high surface area, microporous-dominated structure, and narrow pore size distribution are the best candidate sorbents for hydrogen adsorption. The hydrogen adsorption capacity of activated carbons was investigated with different surface areas (1000–3300 m^2/g) and it was found that the hydrogen storage capacity of carbon materials was proportional to their specific surface area and the volume of micropores; the narrow micropores were preferred for the adsorption of hydrogen [20]. Numerous efforts have been made to develop microporous carbons with ultra-high surface area by chemical activation. It has been demonstrated that activated carbons prepared at a high potassium hydroxide (KOH)/carbon mass ratio (3–4) and activation temperature (1073–1173 K) exhibited a high surface area up to 3800 m^2/g and great potential for hydrogen adsorption. Table 8.1 summarizes the hydrogen adsorption capacities of activated carbons prepared

TABLE 8.1
Hydrogen Adsorption Capacity of Some Carbonaceous Nanomaterials Reported

Sample	Activating Agent	S_{BET} (m^2/g)	H_2 Uptake (wt.%)	Conditions T/P (K/MPa)	References
Activated carbons					
Chitosan – C-PC-2	KOH	3290	2.8 (6.9)	77/0.1(2)	Huang et al. [23]
Carbon dots	KOH	3073	2.4	77/0.1	Wang et al. [22]
Neolamarckia cadamba	KOH	3246	2.8	77/0.1	Hu et al. [24]
Polyacrylonitrile	KOH	2564	2.7	77/0.1	Li et al. [25]
Sawdust	KOH	3477	3.6 (6.7)	77/0.1(2)	Balahmar et al. [26]
Wood chips	KOH	2835	6.30	77/2.0	Pedicini et al. [27]
Cellulose acetate	KOH	3771	3.9 (8.1)	77/0.1(2)	Blankenship et al. [21]
Carbon nanotubes					
SWCNTs	—	—	3.5–4.5	298 /0.04	Liu et al. [28]
MWCNTs	—	—	2	293 /6.5	
Others					
Carbon nanofibers	—	818	2.3	77/1	Vergara-Rubio et al. [29]
Ni doped carbon nanofibers	KOH	3058	5.1	77/10	Hwang et al. [30]
Carbon aerogel	—	3200	5.3	77/ 20	Kabbour et al. [31]

using KOH as an activating agent and biowaste as a precursor. The hydrogen adsorption capacity could reach up to 4.14 wt.% and 8.10 wt.% at 77 K, 0.1 and 20 MPa of H_2, respectively. In addition to surface area and micropores, heteroatom groups on carbon surfaces were also found to play a vital role in hydrogen adsorption. The oxygen-rich surface created by oxygen functional groups (COOH, C–OH and O–C=O) could effectively enhance gravimetric hydrogen storage [21]. Nitrogen and sulfur-doped porous carbon, as investigated by Wang et al. [22], exhibits sulfonyl groups on its surface. These groups exhibit the greatest affinity for H2 molecule adsorption, resulting in a substantial heat of adsorption of 9 kJmol−1. In contrast, activated carbons typically achieve only 4–5 kJmol^{-1}, as reported in the same study by Wang et al. [22]. Table 8.1 illustrates the adsorption capacity of some carbonaceous nanomaterials.

Due to the nature of physical adsorption, hydrogen adsorption on activated carbons is sensitive to pressure and temperature. Although many activated carbons exhibit high hydrogen adsorption capacity at cryogenic temperature and even low hydrogen partial pressure, only small quantities of hydrogen could be adsorbed at room temperature and moderate pressure. Metal-nanoparticle (Ni, Pt, Pd) doping is a promising strategy to enhance hydrogen uptake at room temperature. The doped metal nanoparticles could help dissociate the hydrogen molecules and lead to spill over of the hydrogen atoms into the porous material. The hydrogen storage capacity of activated carbons depends linearly on Pd content at 298 K and hydrogen pressures up to 1 MPa [32].

8.2.2 CARBON NANOTUBES (CNTs)

Carbon nanotubes (CNTs) first discovered by Iijima (1991) [33] are a group of nanomaterials that are formed by a rolled graphene layer with an inner diameter ranging from 0.7 to up to 50 nm and a length of 10–100 μm. Carbon nanotubes (CNTs) can be categorized into three groups: armchair, zigzag, and chiral, depending on how the graphene sheet is folded into a tube structure. Moreover, CNTs can also be classified as single-walled carbon nanotubes (SWCNTs) or multi-walled carbon nanotubes (MWCNTs) based on the number of graphene layers. Owing to their unique electrical, mechanical, and thermal properties, CNTs have been widely used in many low-carbon energy technologies such as fuel cell designs, photovoltaics, biomedicine, and hydrogen adsorption. Several techniques have been developed to produce CNTs including arc discharge, laser ablation, and chemical vapor deposition (CVD). Among these, CVD has emerged as a promising technology to produce CNTs due to their capability of controlling the growth direction of CNTs and scalability [33]. A schematic of CNT preparation methods is presented in Figure 8.2.

Despite their relatively lower surface area and pore volume than activated carbons, CNTs exhibit higher hydrogen adsorption capacities especially at ambient temperature and moderate pressure (see Table 8.1), suggesting the presence of a stronger interaction force between hydrogen molecules and CNTs. Molecular simulation methods have been used to analyze hydrogen adsorption on CNTs and the results demonstrated that the interaction force between hydrogen and CNTs varied with the

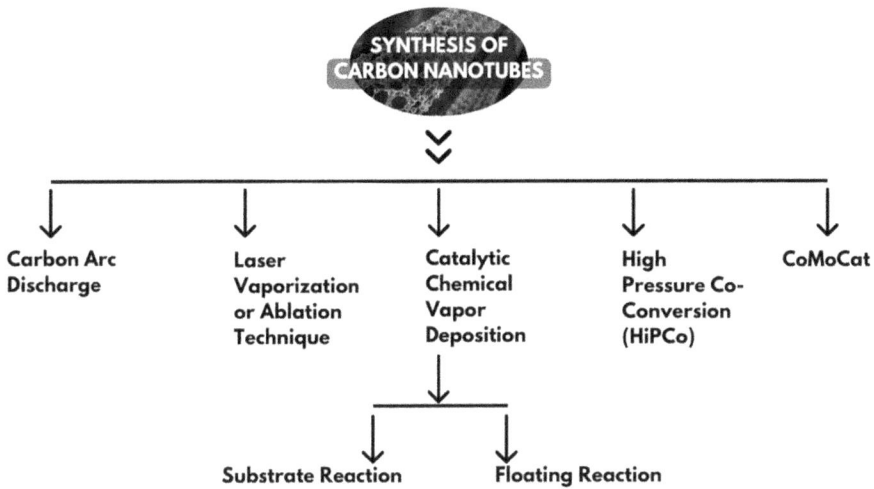

FIGURE 8.2   Schematics of carbon nanotubes preparation methods.

different adsorption sites on CNTs. The adsorption potential of different sites is in the following order: interstices > the endohedral sites along the inner cylindrical surface > the endohedral sites along the outer cylindrical surface. The interaction between hydrogen and the inner tube cavity is stronger than the planar surface and slit pores of similar size. Calculated by Wang and Johnson [34], the adsorption potential of a (9,9) tube array was much higher than carbon split pores with a pore width of 0.9 nm. With an increasing diameter of the carbon nanotube, the adsorption potential decreases and interstitial adsorption takes a more significant role in hydrogen adsorption.

Numerous investigations have demonstrated that CNTs exhibit high hydrogen adsorption capacity of 3.50–6.30 wt.% at ambient temperature and H_2 pressure of up to 14.8 MPa, which is much higher than activated carbons. In addition to the operating conditions such as adsorption temperature, and hydrogen pressure, the physicochemical properties of CNTs play a critical role in hydrogen adsorption. Lee et al. [35] found that the hydrogen adsorption capacity of CNTs increased linearly with increasing tube diameter of SWCNTs. However, only (5,5, 10,10) CNTs were used in their study. To further verify the impact of tube diameter on hydrogen adsorption capacity, Cheng et al. [36] simulated hydrogen adsorption on SWCNTs with tube diameters up to 10 nm by the grand canonical Monte Carlo (GCMC) method.

It was discovered that there exists an optimal diameter of approximately 6 nm for hydrogen physisorption in SWCNTs, rather than a linear increase in hydrogen adsorption capacity with tube diameter. As for MWCNTs, the distance between the walls of MWCNT is also important. Additionally, the defects in CNTs have significant contributions to the hydrogen adsorption capacity. Gayathri et al [37] reported that the adsorption binding energy increased by about 50.50% in the presence of structural defects (e.g., pentagon-octagon defects). Many studies have also been carried out to improve the hydrogen adsorption capacity of CNTs, and metal doping is one of the most widely used strategies. Various alkali metals (Li, K) or metal alloys have been predominantly used for this purpose. For example, it was found that potassium-doped CNTs exhibited high hydrogen adsorption at ambient temperature whereas lithium-doped CNTs showed a high hydrogen adsorption capacity of up to 2.50 wt.% at temperatures as high as 673 K [38].

8.2.3 GRAPHITE

Graphite is a form of carbon wherein all carbon atoms are joined together and arranged in layers. Graphite is inexpensive and can be found in metamorphic and igneous rocks, which can potentially be used for hydrogen storage in large-scale applications. However, the hydrogen adsorption capacity of natural graphite is very low due to its small interlayer distances of only 0.335 nm – even smaller for intercalation of H_2 molecules (0.406 nm) [39] – and low surface area. Expanding the interlayer distance of graphite by using well-defined spacers is an effective strategy to improve its hydrogen adsorption capacity. By introducing ether-solvated alkali metal ions, the spacing between intercalated graphene layers can be expanded from 0.34 nm to 0.46–1.2 nm, depending on the type of ether used [40]. The spacing of THF-solvated Li/K could reach 0.46 nm, while it can sharply increase to 1.20 nm when bidentate ether, DBE, is used. The intercalated alkali metals can not only increase the spacing between graphene layers but also act as acidic cores that can enhance hydrogen adsorption capacity. Li-pillared graphene sheet system exhibited a high hydrogen adsorption capacity of 6.50 wt.% at ambient temperature and 2 MPa of hydrogen, which satisfies the Department of Energy (DOE) target of hydrogen-storage materials for transportation [31]. In addition to changing the spacing, ball-milling of graphite increases its surface area by reducing particle size – resulting in high hydrogen adsorption capacity of 7.40 wt.% [42].

8.2.4 OTHERS

In addition to the carbonaceous materials, carbon nanofibers, carbon aerogel, and spherical fullerenes C60 have also been tested as sorbents for hydrogen adsorption. Carbon nanofibers are a stack of graphene sheets, which are fixed in several alignments to the fiber axis. Similar to activated carbon, the chemical activation method has been widely used to increase the surface area and pore volume of carbon nanofibers. The hydrogen adsorption capacity is proportional to the surface area of carbon nanofibers. Carbon aerogel, one of the lightest carbon materials with a 3D hierarchically micro-meso-macro porous network, has received widespread attention due to its high surface area and pore volume, which has also been tested as hydrogen sorbents. The hydrogen adsorption capacity of carbon aerogel with a surface area over 3000 m^2/g could reach 5.3 wt.% at 77 K and 2 MPa of hydrogen [31]. Similar to carbon nanofibers, most of the studies were carried out in cryogenic conditions, whereas the hydrogen adsorption capacity at ambient temperature was very low (see Table 8.1).

8.3 METAL AND COMPLEX HYDRIDES

8.3.1 METAL HYDRIDES

Magnesium (Mg) is a promising candidate for stationary hydrogen storage which can form a hydride, MgH_2 (with a nominal 7.6 wt.% of hydrogen) and a complex hydride Mg_2FeH_6, $Mg(BH_4)_2$ (with a very high theoretical capacity about 14.8 wt.% of hydrogen). However, at moderate hydrogen pressure, the only phase existing in equilibrium with Mg is magnesium hydride. The main disadvantages of MgH_2 as hydrogen storage material include the high temperature requirement for hydrogenation and dehydrogenation, slow reaction kinetics, and high reactivity towards air and oxygen. The high thermodynamic stability of MgH_2 results in a relatively high desorption enthalpy, which corresponds to an unfavorable desorption temperature of 300°C at 1 bar H_2 [43, 44]. In addition, oxide layers on the metal surface are normally impermeable to hydrogen molecules, preventing hydrogen molecules from transporting into the materials. In magnesium, hydrogenation is limited by the growth of the hydride phase which is controlled by the diffusion of the hydrogen into the particle. Magnesium nanoparticles can alleviate this kinetic rate-limiting step due to their significantly increased surface-to-volume ratio. Additionally, magnesium nanoparticles exhibit many defects, which serve as nucleation sites for the formation of the hydride phase and create large grain boundaries that enhance the diffusion of hydrogen within the matrix.

In 1995, the hydrogen absorption and desorption kinetics in magnesium were improved by mechanically milling magnesium hydride [45], which is a brittle material and therefore easier to mill than pure magnesium [43, 46–48]. Hydrogen absorption rates for ball-milled magnesium samples with various grain sizes (e.g., 1 μm, 50 nm, 30 nm) were studied at 300°C. The nano-crystalline samples absorbed hydrogen rapidly (5.2 and 4.2 wt.% H_2 in 30 nm and 50 nm samples, respectively, in 40 min), in their first cycle without any prior activation, in contrast to the polycrystalline sample (less than 0.5 wt.% H_2 was absorbed) [43].

Further improvement was achieved by milling MgH_2 with the transition metals or their oxides. For example, it was shown that hydrogen could be released from metal oxides milled with nanocrystalline Mg-based hydrides at temperatures as low as 250°C in less than one hour [49–54]. Milling MgH_2 with 1 mol% of TiO_2, V_2O_5, Mn_2O_3, Fe_3O_4 or Nb_2O_5 leads to the same improvement of magnesium hydrogenation and dehydrogenation properties but with the superior catalytic properties of Nb_2O_5 [55]. The real effects of these metal oxides on the kinetic properties of MgH_2 are not known. It is believed that they catalyze the reaction of hydrogen dissociation and recombination at the surface of the magnesium particles and thus accelerate the gas-solid reaction [55]. However, several reports mentioned the role of metal oxide in facilitating the refinement of MgH_2 particle size during milling. Hence, metal oxides could prevent MgH_2 particle agglomeration during mechanical milling which in turn would favor the synthesis of very small magnesium hydride particles with a nano metric size [56, 57]. Smaller particles of magnesium absorb hydrogen faster because diffusion paths are shorter and decompose at lower temperatures compared to larger particles. Milling brittle materials usually leads to fragmentation and a continuous reduction of the particle size until agglomeration prevents further decrease and/or the particles behave in a ductile fashion and cold weld. To avoid agglomeration and cold welding, dispersing agents such as polar, non-polar organic solvents or polymers are added to the grinding medium.

When MgH_2 is milled, the fracturing of MgH_2 particles leads to partial release of hydrogen and a strong agglomeration happens after 20 h of milling. Metal oxides could act as process control agents and prevent strong agglomeration of MgH_2, facilitating the stabilization of nanoparticles and thus avoiding cold welding [57, 58]. A perfect example of this is MgO, which once milled with MgH_2, leads to a material that shows hydrogen kinetics as good as obtained for magnesium modified with transition metals [57]. Using oxides with strong ionic character (polar-oxides i.e. Al_2O_3, TiO_2) reduces the problem of agglomeration during mechanical milling as compared to metal oxides with a weak ionic character (near covalent bond), such as SiO_2. Furthermore, during cycling, the presence of the metal oxides could prevent the sintering of the magnesium nanoparticles [57, 58]. The dehydrogenation properties of α-phase AlH_3 were successfully improved by simultaneous ball milling and doping [59, 60]. Ball milling led to a decrease in the dehydrogenation temperature from 175°C–200°C to 125°C–175°C. Ball milling with alkali metal hydrides such as LiH, NaH, and KH is effective for reducing the dehydrogenation temperature by forming new phases on the surface of AlH_3, which mitigates the effect of Al_2O_3 by creating a window for dehydrogenation [61, 62].

AB alloys are also promising candidates for stationary hydrogen storage applications since gravimetric capacity is less important compared to mobile applications. TiFe is the most well-known alloy in this class, and it is particularly appealing because of its reasonable maximum hydrogen storage capacity of approximately 1.9 wt.% at room temperature [63], low raw material cost and moderate conditions for reversible hydrogen absorption/desorption. Nevertheless, TiFe possesses several disadvantages. The practical utilization of TiFe, manufactured using traditional metallurgical techniques, is hindered by its complex and time-consuming activation process [63]. It also exhibits sensitivity to gaseous impurities in hydrogen after the activation procedure [64]. Nanostructuring proved to be an effective method to overcome the activation problem, by allowing TiFe nanoparticles to instantly absorb hydrogen at room temperature without the need for activation.

8.3.2 COMPLEX HYDRIDES

The introduction of complex hydrides to hydrogen storage research has undoubtedly broadened the materials' scope and enhanced researchers' capabilities to design and optimize new materials. These hydrides are composed of an anionic metal-hydrogen complex, or nonmetal-hydrogen complex bonded to the cationic alkali or transition metal. Thus, the entire large group can be subdivided into two categories. Group I and group II salts of $[AlH_4]^-$, $[NH_2]^-$, $[BH_4]^-$, that is alanates, amides and borohydrides, and transition metal complex hydrides that have anionic (TMHx)-complexes attached to the cationic light metal, such as Mg_2FeH_6 [65]. Several attempts have been made to improve the hydrogen sorption properties of these systems by mechanical milling [66–68]. For example, hydrogen storage properties of catalyzed $NaAlH_4$ and Na_3AlH_6 were evaluated by mechanical milling with liquid alkoxides Ti $(Obun)_4$, $Zr(OPri)_4$ and dry milling with $TiCl_3$. The alkoxides-catalyzed materials showed low reversible capacities and released significant amounts of hydrocarbons upon dehydrogenation. However, these problems were mitigated by a $TiCl_3$ catalyst, introduced during mechanical milling. The reversible hydrogen storage properties of Ti-catalyzed $NaAlH_4$ were studied as a function of Ti-content by ball milling of $NaAlH_4$ and $TiCl_3$ mixtures. It was argued that zero-valent Ti works as a catalyst and $TiCl_3$ is only the precursor [69]. $NaAlH_4$ was produced via a mechano-chemical synthesis procedure [70] and after 2 h of ball milling undoped $NaAlH_4$ could desorb ~3 wt.% H_2 within 2 h at 160°C. The effects of mechanical ball milling on the micro-structural characteristics of $NaAlH_4$ were investigated in the presence of catalysts and observed fracture and fragmentation of particles [71]. Moreover, the hydrogen storage performance of $NaAlH_4$ was found to be highly dependent on the milling time which increases with time [72]. Carbon has also been milled together with complex hydrides as an activator. It was shown that a mixture of $NaAlH_4$ with carbon is capable of reversible hydrogenation and dehydrogenation at much lower pressure and with much faster kinetics than conventional compounds [73]. Ball milling and catalyst addition to $LiAlH_4$ have also been extensively studied. The main catalysts studied being Ti, $TiCl_4$, $TiCl_3$, $TiCl_3$-1/3$AlCl_3$, VCl_3 and $FeCl_3$. However, mechanical milling of metal chlorides with $LiAlH_4$ was complicated because these dopants facilitate decomposition of $LiAlH_4$ even at short milling times, therefore, strict control of milling conditions was required [74–76].

The effect of ball milling on the two-step dehydrogenation reaction of $LiAlH_4$ (Equations 9.1-9.2) was investigated [77]. Ball milling can only increase the first-step reaction rate by decreasing the grain size and has virtually no effect on the dehydrogenation of the second step. The first step reaction in the solid state is limited by a mass transfer process such as long-range atomic diffusion of Al, while the second step of the reaction is controlled by intrinsic kinetics (too low temperature). Therefore, ball milling alone is typically not sufficient to improve the kinetics for the second step of the reaction and a suitable catalytic additive should be used.

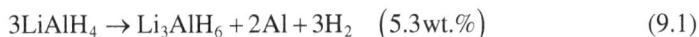

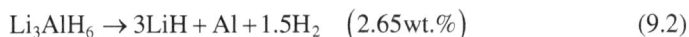

$$3LiAlH_4 \rightarrow Li_3AlH_6 + 2Al + 3H_2 \quad (5.3 \text{wt.\%}) \qquad (9.1)$$

$$Li_3AlH_6 \rightarrow 3LiH + Al + 1.5H_2 \quad (2.65 \text{wt.\%}) \qquad (9.2)$$

The effect of various chloride catalysts on the hydrogen desorption characteristic of $LiAlH_4$ was investigated by Kojima et al. [78]. The catalytic activity, defined by the hydrogen desorption capacity, decreased according to the series $TiCl_3 > ZrCl_4 > VCl_3 > NiCl_2 > ZnCl_2$, (prepared by mechanochemical reaction) [79]. Resan et al. [79] also reported that the addition of $TiCl_3$ and $TiCl_4$ to $LiAlH_4$ eliminates the first step of the reaction and significantly lowers the desorption temperature of the second step.

Another important issue corresponding to $LiAlH_4$ is the reversibility. Subjecting $TiCl_3$-doped $LiAlH_4$ to 8 MPa of hydrogen pressure at 125°C did not result in successful hydrogenation [80]. Rehydrogenation studies were also carried out for $LiAlH_4$ doped with 5 mol% $LaCl_3$, $Ce(SO_4)_2$, Ti and Ni under ~ 8 MPa of hydrogen pressure at 180°C for 2 h. Only a small amount of hydrogen was absorbed by forming Li_3AlH_6 (0.83 wt.% for Ti, 0.62 wt.% for $LaCl_3$ and 0.12 wt.% for $Ce(SO_4)_2$)) [81]. These results have caused the reversibility of $LiAlH_4$ to be followed by other methods such as mechanical reversibility where rehydrogenation occurs during mechanical milling under the hydrogen atmosphere in the presence of catalysts [82]. A five-step physiochemical pathway for the cyclic dehydrogenation and rehydrogenation of $LiAlH_4$ from Li_3AlH_6, LiH, and Al was developed [83]. $LiBH_4$ could be destabilized by partial cation substitution having a smaller size and/or higher valence state and larger electronegativity than Li [84]. The milling of $LiBH_4$ with MgH_2 led to a material that had a reversible capacity of 8 wt.% of hydrogen under 100 bar at 350°C–450°C [85]. Less success was gained by mixing $LiBH_4$ with $LiNH_2$ because hydrogen is released above 300°C and Li_3BN_2 is formed, which cannot reabsorb hydrogen [86].

8.3.3 NANOCONFINED COMPLEX HYDRIDES

The nanoconfinement approach is a promising method to stabilize the nanoparticles' influence on the thermodynamics and kinetics of hydrides [87]. Nanoconfined MgH_2 was produced by mechanical milling with nanoporous graphite [88]. A stable hydrogen storage capacity for the produced sample was reported after ten hydrogenation and dehydrogenation cycles. Nanocrystals of MgH_2 were embedded in the LiCl matrix following mechanochemical metathesis reaction between LiH and $MgCl_2$. A 12% increase in the equilibrium pressure at 330°C was reported for this sample compared to the bulk MgH_2 [89]. Although nanoconfinement mitigates agglomeration, only small changes in the thermodynamic stability of MgH_2 were reported for particles $\geq$ 5 nm. A nanoconfinement method was also used to lower the dehydrogenation reaction temperature in the $LiBH_4/MgH_2$ mixture [90]. It was shown that the dehydrogenation profile of the nanoconfined $2LiBH_4+MgH_2$ mixture could be significantly enhanced while the nanoconfined sample retains significantly higher reaction kinetics at the end of the fourth hydrogen cycle with no indication of any release of borane gases [91]. The hydrogen storage characteristics of bulk and nanoconfined in CO_2-activated carbon aerogel scaffold (30 nm pore size) of $0.7LiBH_4 + 0.3Ca(BH_4)_2$ mixture were investigated. The hydrogen storage capacity of the bulk sample decreased to 7.9 wt.% compared to 8.2 wt.% for the nonconfined mixture after three cycles. A substantial 30% reduction in the dehydrogenation enthalpy of the nonconfined mixture was reported compared to the bulk alloy [92].

8.4 METAL-ORGANIC FRAMEWORKS (MOFs)

Metal-Organic Frameworks (MOFs) have long been regarded as one of the prime contenders to enable high-performance hydrogen storage media. The first example of MOF-based hydrogen storage was shown by Rowsell and Yaghi et al. [91], via the synthesis of MOF-5. Furthermore, this marked the original demonstration of permanent porosity being maintained within the framework material after thermal activation, with the resulting Brunauer-Emmett-Teller (BET) surface area determined to be 2500–3000 m^2 g^{-1}. Hydrogen adsorption measurements revealed the gravimetric uptake to be 4.5 wt.% at 77 K and 20 bar pressure, while remarkably, 1 wt.% was achieved at ambient temperature. This study also provided a preliminary demonstration of the impact of MOF tunability, showing simple linker substituted derivates of

MOF-5 (IRMOF-6 and IRMOF-8) possessed greatly improved hydrogen uptakes at room temperature and 10 bar. These promising uptakes from easily synthesized and manipulated, low-cost materials sparked intensive research to further explore this exciting class of nanomaterials [92].

The MOF hydrogen storage materials can be broken down into two main classes, based on their adsorption mechanisms and their surface area or open metal site interactions [93]. The surface area approach utilizes the weakest of the van der Waals forces – London dispersion forces. The magnitude of the resultant attractive force can be influenced by the analyte's polarizability and the distance between the two bodies. Therefore, MOF chemical structure and pore geometry can be optimized to enhance the dispersion forces and the resulting hydrogen uptake [91]. Conversely, the open metal site (OMS) approach works by introducing alternative high-energy metal-binding sites. These unsaturated metal ion sites possess much higher values of isosteric heat of hydrogen adsorption (Q_{st}), in the region of −10 to −20 kJmol^{-1}, when compared to −1 to −10 kJmol^{-1} for porous surface area-based MOFs [94, 95]. These sites are generated by employing metal centers that contain weakly coordinated solvent molecules. The solvent molecules can then be extracted using a simple thermal activation procedure, leaving highly favorable metal binding sites for target hydrogen molecules [96].

8.4.1 CRYOGENIC HYDROGEN STORAGE WITH MOF

The primary focus for MOF hydrogen storage research has been related to cryogenic adsorption. This is due to the relatively low isosteric heat of hydrogen adsorption (Q_{st}), −1 to −10 kJmol^{-1}, possessed by most MOF absorbents. The often reported "Chahine's rule" predicts that hydrogen storage increases at cryogenic temperatures by 1 wt.% for every additional 500 m^2 g^{-1}. However, the improvement drastically decreases to approximately 0.11 wt.% when ambient temperature adsorption is considered Zhao et al. [97]. Consequently, the desire to produce ever more porous frameworks to exploit this storage potential is prevalent within the literature. One such approach to achieve this was using extended linker species, as MOFs linkers can be considered as two or more bonding sites, typically carboxylate or imidazole moieties, separated by an organic carbon structure. By simply extending the carbon component, novel isostructural MOFs can be achieved with enlarged pore sizes. Ultrahigh porosity was observed through the synthesis of MOF-210 and related materials MOF-180, MOF-200 and MOF-205. These frameworks all utilized the $Zn_4O(CO_2)_6$ metal center while coordinated with either one two or extended organic linkers All measured BET surface areas were shown to be approximately 4500 m^2 g^{-1}, with the exception of MOF-210, achieving an impressive 6240 m^2 g^{-1} after supercritical CO_2 activation. This was the highest reported for any crystalline material at the time and translated to an incredible 17.6 wt.% total hydrogen uptake at 77 K and 80 bar [98].

However, the direct lengthening of the linker often generates significant framework instability. Furthermore, the increased pore size within the framework usually leads to structural collapse upon activation. There is also a heightened risk of framework interpenetration, where a secondary framework forms within the expanded pore structure, causing inhibited surface areas and porosity. Copper paddlewheel containing Product Control Number (PCN)-68, which possesses a (3,24)-connected network, enables large expandable mesopores stabilized by fixed micropore windows.

The extended highly branched hexacarboxylic linker was far larger (<11.2 Å) than previously feasible, with other members of the PCN family using the twisted boracite network. This enlarged pore volume resulted in impressive BET surface areas of 5109 $m^2 g^{-1}$, which was reported to be the highest value demonstrated within paddle-wheel cluster-based frameworks. Hydrogen adsorption measurement as determined by the total gravimetric uptake was 13 wt.% at 77 K and 100 bar Yuan et al. [99].

Another interpenetration inhibition concept, leading to enhanced surface areas, was shown through the development of the DUT-32 framework. DUT-32 is a binary linker MOF composed of BTCTB and BPDC; these were selected to provide highly contrasting polar and non-polar local chemical environments. This regional disparity throughout the framework prevents additional framework crystallization. The resulting structure contained a combination of mesopores and micropores, with an exceptionally low theoretical crystal density of 0.27 g cm^{-3}. After optimization of the supercritical CO_2 activation procedure, nitrogen adsorption measurement revealed the BET surface to be 6411 $m^2 g^{-1}$. This porosity resulted in a total H_2 uptake at 77 K and 80 bar of 14.21 wt.%, placing this material in line with the high total gravimetric capacities achieved within porous materials [100].

8.4.2 WORKING CAPACITY AND BALANCED ADSORPTION

Intensive materials-discovery research has shifted focus toward a wider variety of capacity metrics, namely working capacity and volumetric capacity, as well as the traditional gravimetric capacity. Gravimetric capacity considers the uptake on a mass basis ($g_{H2} g_{MOF}^{-1}$ or wt.%), whereas volumetric uptake is related to the volume of MOF material ($g_{H2} L_{MOF}^{-1}$). This is an important distinction, as it is argued that volumetric uptake is the more critical parameter for space-limited applications such as vehicular or mobile power supplies but is seldom reported. The working capacity (either gravimetric or volumetric), also referred to as usable or deliverable capacity, is the amount of hydrogen viable for desorption and utilization based on the theoretical operational conditions used within an absorbent tank [101, 102].

MOF-5 is a benchmark material for possessing a balanced adsorption profile, showing an impressive 4.5 wt.% and 31.1 g L^{-1} uptake for gravimetric and volumetric usable capacity, respectively. Interestingly, a significant capacity impact was observed through employing temperature-pressure swing conditions (77 K/100 bar, 160/5 bar). While both material capacities were raised considerably as expected, MOF-5 usable volumetric capacity exceeded IRMOF-20 (9.1 vs 7.8 wt.% and 51.0 vs 51.9 g L^{-1}, IRMOF-20 and MOF-5 experimental values, respectively) – highlighting the current knowledge gap surrounding the influences of system-level operating conditions on material-level capacities [103]. Consequently, Farha and co-workers presented a systematic benchmarking study which involved conducting high-pressure hydrogen adsorption measurements, in both pressure-swing and temperature-pressure swing configurations, using fourteen structurally diverse MOFs. All materials were shown to have exceeded the US Department of Energy 2020 volumetric system target utilizing a temperature-pressure swing approach; however, the trend was constant. This suggested that volumetric usable capacity is linked to the two conflicting variables: volumetric surface area and void fraction. Comparatively, gravimetric usable uptake was shown to have a positive linear correlation with both pore volume and BET surface area, [7].

Despite the number of real materials available within the Cambridge Structural Database (CSD), there is a wide array of proposed but not yet synthetically realized structures that may still provide value towards computational calculations. The utility of computationally directed MOF research extends beyond simulations of capacity values for screening purposes [104]. Through analysis of the multitude of available MOF structural properties, new structure-performance relationships can be ascertained, and novel high-capacity materials rationally designed. Farha and Stoddart developed the new NU-1501-M series (Al and Fe) of ultra-porous frameworks, from the PET-2 linker and hydrated metal (III) chloride salt. Synthesis was inspired by the relation between the product of the gravimetric and volumetric surface areas (GSA $\times$ VSA), the void fraction (VF) and the largest pore diameter (LPD). The relation was observed from computationally screening an expansive MOF database (2800 frameworks). Materials with balanced surface areas were shown to on average possess 0.85 VF and 17.2 Å LPD. NU-1501-M, an extended form of the known NU-1500 series, was designed to have closely aligned characteristics (VF – 0.87, LPD – 18.8 Å). Consequently, NU-1501-Al and NU-1501-Fe were shown to have among the best H_2 absorbents for balanced deliverable capacity under temperature-pressure swing conditions (M = Al 14.0 wt.% and 46.2 g L^{-1}, M = Fe 13.2 wt.% and 45.4 g L^{-1}) [105].

8.4.3   MOFs for Hydrogen Storage at Ambient Temperature

The long-term aspiration for MOF-based hydrogen storage technologies is to be able to operate at higher temperatures, ideally at room temperature. This eliminates many engineering and design challenges associated with heating/cooling performance and the use of cryogenic liquids. A typical approach to improve this is to raise the isosteric heat of adsorption (Q_{st}) to −15 to −20 kJ mol^{-1}, shown to be the optimum region via theoretical calculations [97].

Jaramillo et al. [106] exhibited the new $V_2Cl_{2.8}$(BTDD) framework, which is composed of unsaturated vanadium salts interlinked with BTDD molecules [H2btdd, bis(1H-1,2,3-triazolo[4,5-b],[4′,5′-i])dibenzo[1,4]dioxin]. The mixed-valent metal center processed a high density of exposed vanadium (II) square-pyramidal sites (60%) compared to saturated vanadium (III) (40%), which were shown to provide backbonding interaction with H_2. This additional interaction enhanced the Q_{st} to −21 kJ mol^{-1}. High-pressure hydrogen measurements (298 K and 100 bar) showed 60% gravimetric improvement over benchmark material Ni-MOF-74, with comparable volumetric uptakes. These materials are also shown to provide superior usable volumetric capacities to compressed H_2 over a range of operational conditions [106].

8.4.4   Shaped MOFs for Hydrogen Storage

While materials discovery and structural optimization are at the forefront of MOF research, with new record-breaking frameworks being produced at a rapid rate, these materials will need to be processed into a suitable macroscopic form for practical use within real adsorption systems. Currently, metal-organic frameworks are synthesized as fine crystalline powders; however, this form presents many issues regarding its practical implementation in a fuel tank. Loose powders are inherently difficult to pack, risking poor utilization of tank space and potential contamination concerns [107].

Thermal transport is also compromised, making the thermal management between adsorption/desorption cycles problematic [108]. Furthermore, the small particle sizes found within MOF powders generate large pressure drops within the gas flow across the storage tank [109]. Numerous shaping strategies have been investigated, such as pelletization, extrusion, 3D printing, sol gel-controlled drying, and spray drying [110].

Due to the prolific nature of pellet presses throughout research laboratories and industry, an array of pelletization papers can be seen in the MOF literature. The consensus is that MOF structures are mechanically unstable and prone to amorphization given sufficient application of pressure. The extent of this amorphization tendency is suggested to be MOF-dependent; however, little has been known about controllable parameters to mediate this phenomenon [111–113]. The influence of solvation, crystallite size and pressure upon the pelletization process for HKUST-1 and Ni-MOF-74 shows that solvation control is crucial to generating high-density materials without sacrificing pore structure. Low solvation (50% pore volume) resulted in a 50% loss in porosity compared to 5% for fully saturated MOF [113]. Larger crystals were shown to produce high porosity pellets at low-pressure conditions (9035 psi); however, the density and pressure can be further enhanced without significant loss of porosity through the use of super-saturated MOF pores (>100% pore volume occupied). Similar trends were also observed for Ni-MOF-74 Wang et al. [114].

Another issue faced by MOF materials is that the volumetric total and usable capacities are often calculated from their single crystal density; however, the manufactured MOF powder is known to have significantly lower bulk densities [11, 112–115]. Fairen-Jimenez and colleges recently demonstrated the application of sol-gel controlled drying to HKUST-1, providing an extension to previous works pertaining to ZIF-8 and UiO-66 [116, 117]. HKUST-1 was selected based on synthesis, scalability and volumetric capacities (deliverable and total), from a group of ten high-performance materials suggested by a GCMC screening of a database containing 2932 experimentally synthesized MOFs. High-pressure hydrogen measurements revealed the total volumetric uptake to be 46.7 g L^{-1} (77 K 100 bar) and usable capacity under temperature pressure swing to be 43.3 g L^{-1} (77 K 100 bar/160 K 5 bar); remarkably these values are determined using the bulk densities of the sol-gel monolith. Both results are record-setting for densified materials under these conditions – attributed to the exceptional bulk density (greater than single crystal density) generated by the sol-gel methodology [118].

8.4.5 MOF Synthesis Procedures

To see MOF-based materials and products adopted within industry, the viability of large-scale synthesis needs to be considered. Classical MOF synthesis procedures within research laboratories are small-scale batch syntheses – often employing several day-long reactions with temperatures up to 220°C [119–121]. These high temperatures require the use of high boiling solvents such as dimethylformamide (DMF) or diethylformamide (DEF), which have poor toxicological profiles. Furthermore, DMF is a listed "Substance of Very High Concern" governed by REACH regulations, due to its toxicity and mutagenic properties. These qualities, energy inefficiency, and high personnel risk make traditional solvothermal synthesis inherently unsustainable and not viable to be scaled to industrial quantities. Therefore, to facilitate the industrial-scale production of these materials, the development of optimized synthetic strategies is essential (Table 8.2).

TABLE 8.2

A Comparison of the Physical Form, Surface Area, Total and Usable Capacities under Pressure and Pressure-Temperature Swing Operating Conditions for Selected MOFs

MOF	Macroscopic Form	BET Surface Area m² g⁻¹	Pressure Swing Total Capacity		Pressure Swing Working Capacity		Pressure-Temperature Swing Working Capacity		Reference
			wt.%	g L⁻¹	wt.%	g L⁻¹	wt.%	g L⁻¹	
HKUST-1	Powder	1980	5.3	44.9	2.0	17	5.2	46	(García-Holley et al. [7])
NOTT-112	Powder	3440	8.7	42.7	5.3	24	9.1	41	(Zhao et al. [32])
NU-125	Powder	3230	8.2	51.6	4.1	24	8.5	49	
rht-MOF-7	Powder	1950	4.9	40.5	1.8	14	4.7	37	
Cu-MOF-74	Powder	1270	3.1	43.0	1.0	13	3.0	39	
PCN-250	Powder	1780	5.4	50.8	1.8	16	5.2	47	
NU-1000	Powder	2200	14.1	48.1	5.2	30	8.3	48	
UiO-67	Powder	2360	5.9	42.8	2.9	20	6.0	41	
UiO-68–Ant	Powder	3030	7.6	49.8	4.3	26	7.8	47	
CYCU-3-Al	Powder	2450	8.2	42.9	5.5	27	8.7	41	
Zn2(bdc)2(dabco)	Powder	2020	4.9	44.9	1.6	14	4.8	42	
NU-1101a	Powder	4340	9.5	48.7	6.1	30	9.1	47	
NU-1102a	Powder	3720	9.9	45.3	6.9	31	9.6	44	
NU-1103a	Powder	6245	12.9	44.9	10.1	33	12.6	43	
MFU-4l	Powder	3160	7.7	46.5	4.5	27.5	7.3	44.3	(Zhao et al. [97])
MOF-5	Powder	3512	8.0	53.3	4.5	31.1	7.8	51.9	
NU-1500-Al	Powder	3560	8.4	46.6	4.4	26.6	8.2	44.6	
MFU-4l-Li	Powder	4070	9.9	52.4	5.9	32.4	9.4	50.2	
IRMOF-20	Powder	4073	9.3	52.7	5.7	33.4	9.1	51.0	
BUT-22	Powder	4380	12.0	45.8	7.8	30.3	11.6	44.1	
SNU-70	Powder	4944	10.7	48.9	7.8	34.3	10.6	47.9	
UMCM-9	Powder	5039	11.5	48.5	7.3	34.1	11.3	47.4	
NU-1501-Fe	Powder	7140	13.6	46.9	9.4	33.9	13.2	45.4	
NU-1501-Al	Powder	7310	14.5	47.9	10.3	35.5	14.0	46.2	
HKUST-1	Monolith	1552	4.3	46.7	2.0	21.1	4.0	43.3	(Madden et al. [118])

Consequently, in recent years, the research focus has shifted toward greener and increasingly more scalable approaches [122, 123]. Increasing the space-time yield (STY, kg m^{-3} s^{-1}; that is, the mass of product produced per volume of the reaction mixture in a given period, which is an important chemical engineering parameter that determines the efficiency of a process) is essential. Transitioning from batch to continuous systems is one such method to achieve this, with minimal reactor down times and enhanced automation often resulting in much greater space-time yields. Continuous synthesis of metal-organic frameworks has been demonstrated utilizing a variety of reaction methods: from typical liquid phase systems, solid phase mechanochemical approaches, and, more recently, aerosol spray drying. Each approach is vastly different. Liquid phase systems involve the reaction of precursor flow streams together (often heated by a particular energy source), whereas mechanochemistry utilizes mechanical grinding to cause solid reactants to undergo reaction; spray drying works by rapidly heating a starting material aerosol to form a powdered product [124–126]. The crystallization process is vital to producing metal-organic frameworks with high porosity and yield. Traditional batch synthesis approaches have successfully crystallized many fragile and complex frameworks, due to the use of ideal solvent mixtures and crystallization conditions [71, 127]. Mechanochemical approaches rely on rapid reaction times in little to no solvent, which is ideal from an industrial waste production standpoint and allows the usage of cheap insoluble metal oxide precursors. However, many structurally intricate frameworks will not form due to the harsh conditions and the lack of precise control [128]. Whereas, liquid phase systems, such as continuous solvothermal, microwave or ultrasonication can manipulate the same or related variables found in traditional batch style experiments; this allows greater flexibility in the range of materials, which can be produced and their structural quality. However, these approaches rely on large quantities of bulk solvent, which further increases the process cost and involves additional post-synthesis solid-liquid separation and drying requirements [129]. Spray-drying systems, similar to mechanochemical ones, enable the efficient generation of dried MOF products. Typically, micrometer-sized beads are formed from the rapid drying of precursor aerosols. Industrial-scale systems have inbuilt solvent recovery and recycling to minimize solvent wastage. However, due to the inherent hazards associated with large volumes of aerosolized organic solvents, water needs to be used instead for the process to be feasible. This limits the scope of the process and imposes an additional energy penalty from the high heat capacity of water [130].

8.4.6 Challenges in MOF Applications in Hydrogen Storage

While MOFs have been heralded for nearly 20 years since the pioneering work by the Yaghi research team [91], several challenges still require addressing before seeing MOF-based hydrogen storage technology become a mainstay of the energy sector. These primarily revolve around the cost of manufacture, development, and scaling of shaping strategy, as well as bridging knowledge gaps influencing MOF uptake under different system-level operating conditions. With more than 90,000 experimentally validated MOFs present within the Cambridge Structural Database, we have seen targeted adsorption metrics (surface area and total and usable capacities) increasingly enhanced [131]. However, a minuscule fraction of those available materials have been translated to continuous processes or demonstrated via industrially relevant production scales. Consequently, these inefficient synthetic methods have resulted in prohibitive costs and limited commercial availability. Pioneering materials companies such as Promethean Particles, NovaMOF, MOF Technologies, MOF Apps and BASF, are attempting to change this, but further work needs to be done [132]. Another challenge facing MOFs is the lack of shaping methodology to transform the as-synthesized MOF product into a viable practical form for adsorption processes. While numerous potential techniques have been demonstrated, most fall foul of pore blocking from the inclusion of additional additives or amorphization from mechanical pressure. Furthermore, limited studies have exhibited scaling or a substantive MOF scope, inhibiting the process adoption from a commercial standpoint [133]. Finally, with the recent adoption of temperature-pressure swing as a viable approach to enhance the working capacities of high-performing absorbents, numerous studies have revealed unexpected capacity improvements, with certain frameworks such MOF-5 or Cu-MOF-74 gaining much more benefit from the increased temperatures compared to other materials. More research needs to focus on developing the structural-performance relations for materials under this operating regime [7, 103]. Additionally, one may question whether this disparity translates to high-density shaped MOFs. Unfortunately, few shaped materials have been tested under these conditions.

8.5 COVALENT ORGANIC FRAMEWORKS (COFs)

8.5.1 SYNTHESIS OF COFs

Since the first report of covalent organic frameworks (COFs) [134], a diverse set of monomer structures have been used to synthesize various types of COFs. The initial COF-1 and COF-5 utilized condensation reactions of diboronic acid and hexahydroxytriphenylene and achieved permanent rigid porous architectures with pore sizes ranging from 7 to 27 angstroms and high surface areas. Co-condensation reactions between. [1262,3,6,7,10,11-hexahydroxytriphenylene (HHTP) and 1,3,5-benzenetriboronic acid (BTBA), 1,3,5-benzenetris (4-phenylboronic acid) (BTPA), and 4,4′-biphenyldiboronic acid (BPDA) produced COF-6, -8, and -10, respectively [134, 135]. COFs-6, -8, and -10 crystallize as 0.5–3.0 μm hexagonal platelets. These 2D COFs are periodic in all three dimensions. In 2007, Cote et al. [135] reported a general design strategy for the synthesis and crystallization of micro- and mesoporous crystalline COFs, resulting in the implementations of three-dimensional covalent organic frameworks (3D COFs) [127]. The 3D COFs (termed COF-102, COF-103, COF-105, and COF-108) were synthesized by self-condensation and co-condensation reactions of the rigid molecular building blocks – the tetrahedraltetra(4-dihydroxyborylphenyl)methane (TBPM), and its silane analog (TBPS), and triangular HHTP targeting two nets based on triangular and tetrahedral nodes: ctn and bor [136]. The substitution of the phenylene moieties of COF-102 with diphenyl, triphenyl, naphthalene and pyrene moieties also produces COF-102-2, COF-102-3, COF-102-4, and COF-102-5. 3D COFs provide full accessibility from within the pores to all the edges and faces of the molecular units and maximize the number of adsorption sites and surface area, with more edges from aromatic rings [137]. Microwave heating has been proved as an alternative means to accelerate chemical reactions. The synthesis of 2D COF-5 and 3D COF-102 were achieved in 20 minutes using microwave instead of 72 h in the solvothermal synthesis [138]. Moreover, the BET surface area of COF-5 (2019 m²/g) obtained is slightly higher than solvothermal reactions (1590 m²/g).

8.5.2 Characterization and Simulation of COFs Nanomaterials

Characterization of COF materials to reveal structural regularity, atomic connectivity, porosity, and morphology is more complex compared to that of MOF. Powder X-ray diffraction (PXRD) is widely used to assess the crystalline quality of COF samples by the signal intensity and assisted computational algorithms such as the Pawley method can provide further information on the unit-cell lattice [139]. The atomic connectivity in the COF including the formation of covalent bonds is normally investigated by solid-state nuclear magnetic resonance (NMR) spectroscopy operated at higher magnetic fields [140]. The porosity and surface area can be assessed by gas adsorption-desorption measurements. Scanning electron microscopy (SEM) can also be utilized to observe the morphology of COF materials. X-ray photoelectron spectroscopy (XPS) is helpful in understanding the state of metal ions incorporated into the COF materials. Based on the classical molecular modelling method, computational modelling of the COF materials for hydrogen storage can provide more information on property predictions [141]. For example, Klontzas and co-workers used a multiscale approach starting with ab initio calculations with MP2 calculations to obtain the hydrogen binding sites and carried out classical grand canonical Monte-Carlo simulations under different thermodynamic conditions to predict the hydrogen uptake of COFs [142]. Table 8.3 shows some widely known COF materials, surface structures, densities, and hydrogen adsorption capacities. For COF-108, the gravimetric uptake reached 21 wt. % in 77 K and 100 bar conditions, which is 2 times larger than that of a typical MOF with comparable volumetric properties. Similar studies on COF systems also indicate that hydrogen storage at ambient temperature is practically possible, particularly for metal-doped COFs. It has been theoretically predicted that lithium-doped COFs should possess improved capacity for hydrogen storage, which is attributed to the proposed formation of a dative bond between the H_2 and the Li atom [143, 144]. The calculated hydrogen uptakes for

TABLE 8.3
Examples of COF Materials Properties [145, 146]

COF Material	Composition	Pore (Å)	BET Surface Area (m²/g)	Bulk Density (g/cm³)	H₂ Uptake at 77 K (mg/g)
COF-1	C_3H_2BO	9	711	0.98	14.8
COF-5	$C_9H_4BO_2$	27	1590	0.58	35.8
COF-6	$C_8H_3BO_2$	9	750	1.1	22.6
COF-8	$C_{14}H_7BO_2$	16	1350	0.71	35.0
COF-10	$C_{12}H_6BO_2$	32	1760	0.48	39.2
COF-102	$C_{25}H_{24}B_4O$	11.5	3472	0.41	72.4
COF-103	$C_{24}H_{24}B_4O$	12.5	4210	0.38	70.5

Li-doped COF-108 is 6.73 wt.%, at the ambient temperature and 100 bar, which supersedes those for MOF materials and non-doped COFs.

8.5.3 Hybridization of COFs and MOFs

As an emerging porous frame material platform, a broad variety of MOF/COF hybrid materials have been synthesized, including COFs covalently bonded to the surface functional groups of MOFs (MOF@COF) [147–149], MOFs grown on the surface of COFs (COF@MOF) [139, 140], and bridge reactions between COF and MOF (MOF+COF) [150, 151]. Extensive studies have shown that the allowed complementary utilization of π-conjugated skeletons and nanopores has equipped these hybrid materials with great potential in hydrogen storage.

8.5.4 Other Nanoporous Polymer-based Nanomaterials

Hyper-crosslinked polymers (HCPs) prepared from the lightly crosslinked polymeric precursors are among other nanoporous polymers which offer great potential for hydrogen storage more practically [152]. For instance, a hyper-crosslinked polystyrene material was synthesized by suspension polymerization of vinylbenzyl chloride followed by a Friedel–Crafts-type post-crosslinking and achieved a BET surface area of 1466 m^2/g with 3.04 wt.% hydrogen absorption at 77 K and 15 bar [153]. Similarly, using gel-type and permanently porous poly(divinylbenzene-co-vinylbenzyl chloride) precursor with hyper-crosslinked Davankov-type resins, a H$_2$ uptake of 1.55 wt.% was achieved at 1.2 bar with surface areas up to 1200 m^2/g resin [154]. Another HCP that exhibits a range of useful engineering properties for large-scale applications, such as good thermal stability and solution processability, is a polymer of intrinsic microporosity PIM-1. This material can be processed into several morphologies while remaining highly porous and exhibiting good thermal and mechanical properties; additionally, the adsorption capacity can be enhanced by mixing highly porous fillers into the PIM-1 matrix to form composite films [155]. More examples of HCPs for hydrogen storage are given in Table 8.4.

TABLE 8.4
Examples of HCP Materialsss for Hydrogen Storage

HCPs	BET Surface Area (m^2/g)	H$_2$ Uptake at 77k and 1 Bar (wt.%)	Reference
EOF-1	780	0.94	Rose et al. [156])
EOF-2	1046	1.21	Germain et al. [157])
Aniline	630	0.96	Rochat et al. [158])
PIM1	500–800	0.5–1.1	Ghanem et al. [159])
Trip-PIM	1065	1.65	Ghanem et al. [159])
OFP-3	1159	1.56	Makhseed et al. [160])
PS4TH	971	3.6	Yuan et al. [161])
PS4AC2	1043	3.7	Yuan et al. [161])
PT4AC	762	2.2	Yuan et al. [161])
CMP0	1018	1.4	Hasell et al. [162])

8.6 CONCLUSION

Hydrogen storage remains one of the most challenging aspects of achieving a successful hydrogen economy. To develop a safe and efficient solid-state hydrogen storage unit, a wide range of nanomaterials/nanocomposites have been recently investigated based on *carbonaceous nanomaterials, metal and complex hydrides, metal-organic frameworks,* and *covalent organic frameworks.* In addition to the adsorption conditions (i.e., temperature and pressure), the physicochemical properties of these nanomaterials (i.e., surface area, pore structure, composition, particle size) have a significant impact on the performance of hydrogen storage. Furthermore, MOFs are key materials being investigated as high-performance hydrogen absorbents through physical adsorption, enabling the storage of larger volumes of gas at lower pressures, thereby promoting the use of hydrogen as a green energy source. The MOFs have shown continued promise for cryogenic hydrogen storage, with record-breaking uptakes being published frequently as both the MOF structure and the operational conditions have been refined. The increased ease of production and reduction in commercial cost will further aid in overcoming, arguably, the most difficult challenge of shaping MOFs in practically usable forms. As for the carbonaceous nanomaterials, the physical adsorption of hydrogen over these materials has gained more attention as a result of their large surface areas, rapid adsorption-desorption kinetics and reversibility. However, a high hydrogen adsorption capacity requires relatively lower adsorption temperatures (much lower than ambient temperature). To increase the hydrogen storage performance, the surface of carbonaceous nanomaterials can be functionalized by (1) heteroatoms, or (2) metal catalysts to support the Kubas interactions and hydrogen spillover effects. Unlike carbonaceous nanomaterials, hydrogen storage at an ambient temperature is practically possible, particularly for metal-doped COFs. In addition to physical storage through MOFs and carbonaceous nanomaterials, chemical storage of hydrogen through metal hydrides and complex hydrides has also shown promising results. However, cost-effectiveness, poor reversibility, and lack of cost-effective material synthesis procedures are challenges to be addressed for long-term hydrogen storage applications. Despite the significant fundamental improvement that has been achieved, future efforts are still required to optimize the existing techniques and to explore new media with excellent hydrogen storage performance to achieve high-energy efficiency and economic viability. Novel efforts toward large-scale continuous manufacturing will be required for these new materials, transitioning from laboratory-scale batch synthesis with hazardous and uneconomical solvent usage. Ideally, this should be considered from the materials-discovery stage to enable faster process development. It is also crucial for further work to explore more efficient control strategies for defect sites, morphology and crystal growth, as this will be critical to reaching higher hydrogen storage capacities.

REFERENCES

1. Dillon, A., Jones, K. M., Bekkedahl, T. A., Kiang, C. H., Bethune, D. S., & Heben, M. J. (1997) 'Storage of hydrogen in single-walled carbon nanotubes', *Nature*, 386(6623), 377–379.

2. Langmi, H. W. et al. (2014) 'Hydrogen storage in metal-organic frameworks: A review', *Electrochimica Acta*, 128(2014), pp. 368–392. doi: 10.1016/j.electacta.2013.10.190

3. Marx, S. et al. (2010) 'Tuning functional sites and thermal stability of mixed-linker MOFs based on MIL-53(Al)', *Dalton Transactions*, 39(16), p. 3795. doi: 10.1039/c002483j

4. Stock, N. & Biswas, S. (2012) 'Synthesis of metal-organic frameworks (MOFs): Routes to various MOF topologies, morphologies, and composites', *Chemical Reviews*, 112(2), pp. 933–969. doi: 10.1021/cr200304e

5. Guillerm, V. et al. (2014) 'A supermolecular building approach for the design and construction of metal–organic frameworks', *Chemical Society Reviews*, 43(16), pp. 6141–6172. doi: 10.1039/C4CS00135D

6. Lu, W. et al. (2014) 'Tuning the structure and function of metal-organic frameworks via linker design', *Chemical Society Reviews*, 43(16), pp. 5561–5593. doi: 10.1039/c4cs00003j

7. García-Holley, P. et al. (2018) 'Benchmark study of hydrogen storage in metal–organic frameworks under temperature and pressure swing conditions', *ACS Energy Letters*, 3(3), pp. 748–754. doi: 10.1021/acsenergylett.8b00154

8. Klontzas, E., Tylianakis, E. & Froudakis, G. E. (2011) 'On the enhancement of molecular hydrogen interactions in nanoporous solids for improved hydrogen storage', *The Journal of Physical Chemistry Letters*, 2(14), pp. 1824–1830. doi: 10.1021/jz2005368

9. Sclüth, F., Bogdanović, B. & Felderhoff, M. (2004) 'Light metal hydrides and complex hydrides for hydrogen storage', *Chemical Communications*, (20), pp. 2249–2258. doi: 10.1039/b406522k

10. Vajo, J. J. & Olson, G. L. (2007) 'Hydrogen storage in destabilized chemical systems', *Scripta Materialia*, 56(10), pp. 829–834. doi: 10.1016/j.scriptamat.2007.01.002

11. Lai, Q. et al. (2019) 'How to design hydrogen storage materials? fundamentals, synthesis, and storage tanks', *Advanced Sustainable Systems*, 3(9), pp. 1–64. doi: 10.1002/adsu.201900043

12. Berbe, V. et al. (2007) 'Size effects on the hydrogen storage properties of nanostructured metal hydrides: A review', *International Journal of Energy Research*, 31, pp. 637–663.

13. Schneemann, A. et al. (2018) 'Nanostructured metal hydrides for hydrogen storage', *Chemical Reviews*, 118(22), pp. 10775–10839. doi: 10.1021/acs.chemrev.8b00313

14. Bansal, R. C., & Goyal, M. (2005). *Activated carbon adsorption*. CRC press.

15. Lillo-Ródenas, M. A., Marco-Lozar, J. P., Cazorla-Amorós, D., & Linares-Solano, A. (2007) 'Activated carbons prepared by pyrolysis of mixtures of carbon precursor/alkaline hydroxide', *Journal of Analytical and Applied Pyrolysis*, 80(1), 166–174.

16. Lowell, S., Shields, J. E., Thomas, M. A., & Thommes, M. (2006). *Characterization of Porous solids and Powders: Surface Area, Pore Size and Density* (Vol. 16). Springer Science & Business Media.

17. Rzepka, M., Lamp, P., & De la Casa-Lillo, M. A. (1998) 'Physisorption of hydrogen on microporous carbon and carbon nanotubes', *The Journal of Physical Chemistry B*, 102(52), 10894–10898.

18. Alonso, J. A., Cabria, I., & López, M. J. (2013) 'Simulation of hydrogen storage in porous carbons', *Journal of Materials Research*, 28(4), 589–604.

19. De la Casa-Lillo, M. A., Lamari-Darkrim, F., Cazorla-Amoros, D., & Linares-Solano, A. (2002) 'Hydrogen storage in activated carbons and activated carbon fibers', *The Journal of Physical Chemistry B*, 106(42), 10930–10934.

20. Xu, W. C., Takahashi, K., Matsuo, Y., Hattori, Y., Kumagai, M., Ishiyama, S., … & Iijima, S. (2007) 'Investigation of hydrogen storage capacity of various carbon materials', *International Journal of Hydrogen Energy*, 32(13), 2504–2512.

21. Blankenship, L. S., Balahmar, N., & Mokaya, R. (2017) 'Oxygen-rich microporous carbons with exceptional hydrogen storage capacity', *Nature Communications*, 8(1), 1–12.

22. Wang, D., Shen, Y., Chen, Y., Liu, L., & Zhao, Y. (2019) 'Microwave-assistant preparation of N/S co-doped hierarchical porous carbons for hydrogen adsorption', *Chemical Engineering Journal*, 367, 260–268.

23. Huang, J., Liang, Y., Dong, H., Hu, H., Yu, P., Peng, L., ... & Liu, Y. (2018) 'Revealing contribution of pore size to high hydrogen storage capacity', *International Journal of Hydrogen Energy*, 43(39), 18077–18082.

24. Hu, W., Huang, J., Yu, P., Zheng, M., Xiao, Y., Dong, H., ... & Liu, Y. (2019) 'Hierarchically porous carbon derived from Neolamarckia cadamba for electrochemical capacitance and hydrogen storage', *ACS Sustainable Chemistry & Engineering*, 7(18), 15385–15393.

25. Li, F. et al. (2019) 'Design and syntheses of MOF/COF hybrid materials via postsynthetic covalent modification: An efficient strategy to boost the visible-light-driven photocatalytic performance', *Applied Catalysis B: Environmental*, 243, pp. 621–628.

26. Balahmar, N., & Mokaya, R. (2019) 'Pre-mixed precursors for modulating the porosity of carbons for enhanced hydrogen storage: Towards predicting the activation behaviour of carbonaceous matter', *Journal of Materials Chemistry A*, 7(29), 17466–17479.

27. Pedicini, R., Maisano, S., Chiodo, V., Conte, G., Policicchio, A., & Agostino, R. G. (2020) 'Posidonia oceanica and wood chips activated carbon as interesting materials for hydrogen storage', *International Journal of Hydrogen Energy*, 45(27), 14038–14047.

28. Liu, C., Chen, Y., Wu, C. Z., Xu, S. T., & Cheng, H. M. (2010) 'Hydrogen storage in carbon nanotubes revisited', *Carbon*, 48(2), 452–455.

29. Vergara-Rubio, A., Ribba, L., Picón Borregales, D. E., Sapag, K., Candal, R., & Goyanes, S. (2022). *Ultramicroporous Carbon Nanofibrous Mats for Hydrogen Storage*. ACS Applied Nano Materials.

30. Hwang, S. H., Kim, Y. K., Seo, H. J., Jeong, S. M., Kim, J., & Lim, S. K. (2021) 'The enhanced hydrogen storage capacity of carbon fibers: The effect of hollow porous structure and surface modification', *Nanomaterials*, 11(7), 1830.

31. Kabbour, H., Baumann, T. F., Satcher, J. H., Saulnier, A., & Ahn, C. C. (2006) 'Toward new candidates for hydrogen storage: High-surface-area carbon aerogels', *Chemistry of Materials*, 18(26), 6085–6087.

32. Zhao, W., Fierro, V., Zlotea, C., Izquierdo, M. T., Chevalier-César, C., Latroche, M., & Celzard, A. (2012) 'Activated carbons doped with Pd nanoparticles for hydrogen storage', *International Journal of Hydrogen Energy*, 37(6), 5072–5080.

33. Khare, R. (2005) 'Carbon nanotube-based composites – a review', *Journal of Minerals and Materials Characterization and Engineering*, 4(01), 31.

34. Wang, Q., & Johnson, J. K. (1999) 'Molecular simulation of hydrogen adsorption in single-walled carbon nanotubes and idealized carbon slit pores', *The Journal of Chemical Physics*, 110(1), 577–586.

35. Lee, S. M., Park, K. S., Choi, Y. C., Park, Y. S., Bok, J. M., Bae, D. J., ... & Lee, Y. H. (2000) 'Hydrogen adsorption and storage in carbon nanotubes', *Synthetic Metals*, 113(3), 209–216.

36. Cheng, J., Yuan, X., Zhao, L., Huang, D., Zhao, M., Dai, L., & Ding, R. (2004) 'GCMC simulation of hydrogen physisorption on carbon nanotubes and nanotube arrays', *Carbon*, 42(10), 2019–2024.

37. Gayathri, V., & Geetha, R. (2007) 'Hydrogen adsorption in defected carbon nanotubes', *Adsorption*, 13(1), 53–59.

38. Yang, R. T. (2000) 'Hydrogen storage by alkali-doped carbon nanotubes – Revisited', *Carbon*, 38(4), 623–626.

39. Pinkerton, F. E., Wicke, B. G., Olk, C. H., Tibbetts, G. G., Meisner, G. P., Meyer, M. S., & Herbst, J. F. (2000) 'Thermogravimetric measurement of hydrogen absorption in alkali-modified carbon materials', *The Journal of Physical Chemistry B*, 104(40), 9460–9467.

40. Inagaki, M., & Tanaike, O. (2001) 'Determining factors for the intercalation into carbon materials from organic solutions', *Carbon*, 39(7), 1083–1090.

41. Deng, W. Q., Xu, X., & Goddard, W. A. (2004) 'New alkali doped pillared carbon materials designed to achieve practical reversible hydrogen storage for transportation', *Physical Review Letters*, 92(16), 166103.

42. Schlapbach, L., & Züttel, A. (2011) 'Hydrogen-storage materials for mobile applications', In *Materials for Sustainable Energy: A Collection of Peer-Reviewed Research and Review Articles from Nature Publishing Group* (pp. 265–270).

43. Zaluska, A., Zaluski, L. & Ström-Olsen, J. O. (1999) 'Nanocrystalline magnesium for hydrogen storage', *Journal of Alloys and Compounds*, 288(1–2), pp. 217–225. doi: 10.1016/S0925-8388(99)00073-0

44. Barkhordarian, G., Klassen, T. & Bormann, R. U. (2004) 'Effect of Nb2O5 content on hydrogen reaction kinetics of Mg', *Journal of Alloys and Compounds*, 364(1–2), pp. 242–246. doi: 10.1016/S0925-8388(03)00530-9

45. Schulz R et al. (1995) 'Nanocrystalline materials for hydrogen storage', *Innovation in Metallic Materials*, p. 529.

46. Gross, K. J. et al. (1996) 'Mechanically milled Mg composites for hydrogen storage: The transition to a steady state composition', *Journal of Alloys and Compounds*, 240(1–2), pp. 206–213. doi: 10.1016/0925-8388(96)02261-X

47. Huot, J. et al. (1999) 'Structural study and hydrogen sorption kinetics of ball-milled magnesium hydride', *Journal of Alloys and Compounds*, 293, pp. 495–500. doi: 10.1016/S0925-8388(99)00474-0

48. Zaluska, A., Zaluski, L. & Ström-Olsen, J. O. (1999) 'Synergy of hydrogen sorption in ball-milled hydrides of Mg and Mg2Ni', *Journal of Alloys and Compounds*, 289(1–2), pp. 197–206. doi: 10.1016/S0166-0462(99)00013-7

49. Liang, G. et al. (1999) 'Catalytic effect of transition metals on hydrogen sorption in nanocrystalline ball milled MgH2-Tm (Tm = Ti, V, Mn, Fe and Ni) systems', *Journal of Alloys and Compounds*, 292(1–2), pp. 247–252. doi: 10.1016/S0925-8388(99)00442-9

50. Liang, G. et al. (2000) 'Hydrogen desorption kinetics of a mechanically milled MgH2+5 at.%V nanocomposite', *Journal of Alloys and Compounds*, 305(1–2), pp. 239–245. doi: 10.1016/S0925-8388(00)00708-8

51. Oelerich, W., Klassen, T. & Bormann, R. (2001) 'Metal oxides as catalysts for improved hydrogen sorption in nanocrystalline Mg-based materials', *Journal of Alloys and Compounds*, 315(1–2), pp. 237–242. doi: 10.1016/S0925-8388(00)01284-6

52. Bobet, J. L. et al. (2003) 'Addition of nanosized Cr2O3 to magnesium for improvement of the hydrogen sorption properties', *Journal of Alloys and Compounds*, 351(1–2), pp. 217–221. doi: 10.1016/S0925-8388(02)01030-7

53. Huot, J. et al. (2003) 'Investigation of dehydrogenation mechanism of MgH2-Nb nanocomposites', *Journal of Alloys and Compounds*, 348(1–2), pp. 319–324. doi: 10.1016/S0925-8388(02)00839-3

54. Castro, F. J. & Bobet, J. L. (2004) 'Hydrogen sorption properties of an Mg + WO3 mixture made by reactive mechanical alloying', *Journal of Alloys and Compounds*, 366(1–2), pp. 303–308. doi: 10.1016/S0925-8388(03)00747-3

55. Barkhordarian, G., Klassen, T. & Bormann, R. (2003) 'Fast hydrogen sorption kinetics of nanocrystalline Mg using Nb2O5 as catalyst', *Scripta Materialia*, 49(3), pp. 213–217. doi: 10.1016/S1359-6462(03)00259-8

56. Song, M. Y., Bobet, J. L. & Darriet, B. (2002) 'Improvement in hydrogen sorption properties of Mg by reactive mechanical grinding with Cr2O3, Al2O3 and CeO2', *Journal of Alloys and Compounds*, 340(1–2), pp. 256–262. doi: 10.1016/S0925-8388(02)00019-1

57. Aguey-Zinsou, K. F. et al. (2006) 'Using MgO to improve the (de)hydriding proper-ties of magnesium', *Materials Research Bulletin*, 41(6), pp. 1118–1126. doi: 10.1016/j.materresbull.2005.11.011

58. Ares, J. R. et al. (2007) 'Influence of impurities on the milling process of MgH2', *Journal of Alloys and Compounds*, 434–435(SPEC. ISS.), pp. 729–733. doi: 10.1016/j.jallcom.2006.08.132

59. Sandrock, G. et al. (2005) 'Accelerated thermal decomposition of AlH3 for hydrogen-fueled vehicles', *Applied Physics A: Materials Science and Processing*, 80(4), pp. 687–690. doi: 10.1007/s00339-004-3105-0

60. Sandrock, G. et al. (2006) 'Alkali metal hydride doping of α-AlH3 for enhanced H2 desorption kinetics', *Journal of Alloys and Compounds*, 421(1–2), pp. 185–189. doi: 10.1016/j.jallcom.2005.09.081

61. Graetz, J. & Reilly, J. J. (2005) 'Decomposition kinetics of the AlH 3 polymorphs', *Journal of Physical Chemistry B*, 109(47), pp. 22181–22185. doi: 10.1021/jp0546960

62. Graetz, J. & Reilly, J. J. (2006) 'Thermodynamics of the α, β and γ polymorphs of AlH3', *Journal of Alloys and Compounds*, 424(1–2), pp. 262–265. doi: 10.1016/j.jallcom.2005.11.086

63. Reilly, J. J. & Wiswall, R. H. (1974) 'Formation and properties of iron titanium hydride', *Inorganic Chemistry*, 13(1), pp. 218–222. doi: 10.1021/ic50131a042

64. Liu, H. et al. (2022) 'Effect of oxygen on the hydrogen storage properties of TiFe alloys', *Journal of Energy Storage*, 55(PB), p. 105543. doi: 10.1016/j.est.2022.105543

65. Borislav, B. & Sandrock, G. (2002) 'Catalyzed complex metal hydrides', *MRS Bulletin*, 27, pp. 712–716.

66. Züttel, A. (2004) 'Hydrogen storage methods', *Naturwissenschaften*, 91(4), pp. 157–172. doi: 10.1007/s00114-004-0516-x

67. David, E. (2005) 'An overview of advanced materials for hydrogen storage', *Journal of Materials Processing Technology*, 162–163(SPEC. ISS.), pp. 169–177. doi: 10.1016/j.jmatprotec.2005.02.027

68. Mandal, T. K. & Gregory, D. H. (2009) 'Hydrogen storage materials: Present scenarios and future directions', *Annual Reports on the Progress of Chemistry – Section A*, 105, pp. 21–54. doi: 10.1039/b818951j

69. Sun, D. et al. (2002) 'X-ray diffraction studies of titanium and zirconium doped NaAlH4: Elucidation of doping induced structural changes and their relationship to enhanced hydrogen storage properties', *Journal of Alloys and Compounds*, 337(1–2), pp. L8–L11. doi: 10.1016/S0925-8388(01)01955-7

70. Zaluski, L., Zaluska, A. & Ström-Olsen, J. O. (1999) 'Hydrogenation properties of com-plex alkali metal hydrides fabricated by mechano-chemical synthesis', *Journal of Alloys and Compounds*, 290(1–2), pp. 71–78. doi: 10.1016/S0925-8388(99)00211-X

71. Thomas, G. J. et al. (2002) 'Microstructural characterization of catalyzed NaAlH4', *Journal of Alloys and Compounds*, 330–332, pp. 702–707. doi: 10.1016/S0925-8388(01)01538-9

72. Wang, P. & Jensen, C. M. (2004) 'Method for preparing Ti-doped NaAlH4 using Ti powder: Observation of an unusual reversible dehydrogenation behavior', *Journal of Alloys and Compounds*, 379(1–2), pp. 99–102. doi: 10.1016/j.jallcom.2004.02.006

73. Zaluska, A., Zaluski, L. & Ström-Olsen, J. O. (2000) 'Sodium alanates for reversible hydrogen storage', *Journal of Alloys and Compounds*, 298(1–2), pp. 125–134. doi: 10.1016/S0925-8388(99)00666-0

74. Balema, V. P., Dennis, K. W. & Pecharsky, V. K. (2000) 'Rapid solid-state transfor-mation of tetrahedral [AlH4]– into octahedral [AlH6]3- in lithium aluminohydride', *Chemical Communications*, (17), pp. 1665–1666. doi: 10.1039/b004144k

75. Balema, V. P. et al. (2001) 'Titanium catalyzed solid-state transformations in LiAlH4 during high-energy ball-milling', *Journal of Alloys and Compounds*, 329(1–2), pp. 108–114. doi: 10.1016/S0925-8388(01)01570-5

76. Andreasen, A. (2006) 'Effect of Ti-doping on the dehydrogenation kinetic parameters of lithium aluminum hydride', *Journal of Alloys and Compounds*, 419(1–2), pp. 40–44. doi: 10.1016/j.jallcom.2005.09.067

77. Andreasen, A., Vegge, T. & Pedersen, A. S. (2005) 'Dehydrogenation kinetics of as-received and ball-milled LiAlH4', *Journal of Solid State Chemistry*, 178(12), pp. 3672–3678. doi: 10.1016/j.jssc.2005.09.027

78. Kojima, Y. et al. (2008) 'Hydrogen release of catalyzed lithium aluminum hydride by a mechanochemical reaction', *Journal of Alloys and Compounds*, 462(1–2), pp. 275–278. doi: 10.1016/j.jallcom.2007.08.015

79. Resan, M. et al. (2005) 'Effects of various catalysts on hydrogen release and uptake characteristics of LiAlH4', *International Journal of Hydrogen Energy*, 30(13–14), pp. 1413–1416. doi: 10.1016/j.ijhydene.2004.12.009

80. Ebner, A. D. & Ritter, J. A. (2005) 'On the reversibility of hydrogen storage in Ti- and Nb-catalyzed Ca(BH4)2', *Adsorption*, 11, pp. 811–816. doi: 10.1016/j.jpowsour.2008.02.094

81. Xueping, Z. & Shenglin, L. (2009) 'Study on hydrogen storage properties of LiAlH4', *Journal of Alloys and Compounds*, 481(1–2), pp. 761–763. doi: 10.1016/j.jallcom.2009.03.089.

82. Wronski, Z. et al. (2007) 'Mechanochemical synthesis of nanostructured chemical hydrides in hydrogen alloying mills', *Journal of Alloys and Compounds*, 434–435(SPEC. ISS.), pp. 743–746. doi: 10.1016/j.jallcom.2006.08.301

83. Wang, J., Ebner, A. D. & Ritter, J. A. (2006) 'Physiochemical pathway for cyclic dehydrogenation and rehydrogenation of LiAlH4', *Journal of the American Chemical Society*, 128(17), pp. 5949–5954. doi: 10.1021/ja0600451

84. Nakamori, Y. & Orimo, S. ichi (2004) 'Destabilization of Li-based complex hydrides', *Journal of Alloys and Compounds*, 370(1–2), pp. 271–275. doi: 10.1016/j.jallcom.2003.08.089

85. Vajo, J. J., Skeith, S. L. & Mertens, F. (2005) 'Reversible storage of hydrogen in desta-bilized LiBH 4', *Journal of Physical Chemistry B*, 109(9), pp. 3719–3722. doi: 10.1021/jp040769o

86. Pinkerton, F. E. et al. (2005) 'Hydrogen desorption exceeding ten weight percent from the new quaternary hydride Li 3BN 2H 8', *Journal of Physical Chemistry B*, 109(1), pp. 6–8. doi: 10.1021/jp0455475

87. Gross, A. F. et al. (2009) 'Fabrication and hydrogen sorption behaviour of nanopar-ticulate MgH 2 incorporated in a porous carbon host', *Nanotechnology*, 20(20). doi: 10.1088/0957-4484/20/20/204005

88. Paskevicius, M., Sheppard, D. A. & Buckley, C. E. (2010) 'Thermodynamic changes in mechanochemically synthesized magnesium hydride nanoparticles', *Journal of the American Chemical Society*, 132(14), pp. 5077–5083. doi: 10.1021/ja908398u

89. Nielsen, T. K. et al. (2010) 'Article Reaction', 4(7), pp. 3903–3908.

90. Javadian, P. et al. (2015) 'Hydrogen storage properties of nanoconfined LiBH4-Ca(BH4)2', *Nano Energy*, 11, pp. 96–103. doi: 10.1016/j.nanoen.2014.09.035

91. Rowsell, J. L. C. & Yaghi, O. M. (2005) 'Strategies for hydrogen storage in metal-organic frameworks', *Angewandte Chemie - International Edition*, 44(30), pp. 4670–4679. doi: 10.1002/anie.200462786

92. Rosi, N. L. et al. (2003) 'Hydrogen storage in microporous metal-organic frameworks', *Science*, 300(5622), pp. 1127–1129. doi: 10.1126/science.1083440

93. Suh, M. P. et al. (2012) 'Hydrogen storage in metal-organic frameworks', *Chemical Reviews*, 112(2), pp. 782–835. doi: 10.1021/cr200274s

94. Vitillo, J. G. et al. (2008) 'Role of exposed metal sites in hydrogen storage in MOFs', *Journal of the American Chemical Society*, 130(26), pp. 8386–8396. doi: 10.1021/ja8007159

95. Bae, Y.-S. & Snurr, R. Q. (2010) 'Optimal isosteric heat of adsorption for hydrogen storage and delivery using metal–organic frameworks', *Microporous and Mesoporous Materials*, 132(1–2), pp. 300–303. doi: 10.1016/j.micromeso.2010.02.023

96. Liu, Y. et al. (2008) 'Increasing the density of adsorbed hydrogen with coordinatively unsaturated metal centers in metal-organic frameworks', *Langmuir*, 24(9), pp. 4772–4777. doi: 10.1021/la703864a

97. Zhao, D. et al. (2022) 'Porous metal–organic frameworks for hydrogen storage', *Chemical Communications*, 58(79), pp. 11059–11078. doi: 10.1039/D2CC04036K

98. Furukawa, H. et al. (2010) 'Ultrahigh porosity in metal-organic frameworks', *Science*, 329(5990), pp. 424–428. doi: 10.1126/science.1192160

99. Yuan, D. et al. (2010) 'An isoreticular series of metal-organic frameworks with dendritic hexacarboxylate ligands and exceptionally high gas-uptake capacity', *Angewandte Chemie International Edition*, 49(31), pp. 5357–5361. doi: 10.1002/anie.201001009

100. Grünker, R. et al. (2014) 'A new metal-organic framework with ultra-high surface area', *Chemical Communications*, 50(26), p. 3450. doi: 10.1039/c4cc00113c

101. Gómez-Gualdrón, D. A. et al. (2017) 'Understanding volumetric and gravimetric hydrogen adsorption trade-off in metal-organic frameworks', *ACS Applied Materials and Interfaces*, 9(39), pp. 33419–33428. doi: 10.1021/acsami.7b01190

102. Broom, D. P. et al. (2019) 'Concepts for improving hydrogen storage in nanoporous materials', *International Journal of Hydrogen Energy*, 44(15), pp. 7768–7779. doi: 10.1016/j.ijhydene.2019.01.224

103. Ahmed, A. et al. (2017) 'Balancing gravimetric and volumetric hydrogen density in MOFs', *Energy and Environmental Science*, 10(11), pp. 2459–2471. doi: 10.1039/c7ee02477k

104. Ahmed, A. et al. (2019) 'Exceptional hydrogen storage achieved by screening nearly half a million metal-organic frameworks', *Nature Communications*, 10(1), p. 1568. doi: 10.1038/s41467-019-09365-w

105. Chen, Z. et al. (2020) 'Balancing volumetric and gravimetric uptake in highly porous materials for clean energy', *Science*, 368(6488), pp. 297–303. doi: 10.1126/science.aaz8881

106. Jaramillo, D. E. et al. (2021) 'Ambient-temperature hydrogen storage via vanadium(ii)-dihydrogen complexation in a metal–organic framework', *Journal of the American Chemical Society*, 143(16), pp. 6248–6256. doi: 10.1021/jacs.1c01883

107. Ren, J. et al. (2015) 'Review on processing of metal-organic framework (MOF) materials towards system integration for hydrogen storage', *International Journal of Energy Research*, 39(5), pp. 607–620. doi: 10.1002/er.3255

108. Bénard, P. & Chahine, R. (2007) 'Storage of hydrogen by physisorption on carbon and nanostructured materials', *Scripta Materialia*, 56(10), pp. 803–808. doi: 10.1016/j.scriptamat.2007.01.008

109. Bueken, B. et al. (2017) 'Gel-based morphological design of zirconium metal-organic frameworks', *Chemical Science*, 8(5), pp. 3939–3948. doi: 10.1039/c6sc05602d

110. Valizadeh, B., Nguyen, T. N. & Stylianou, K. C. (2018) 'Shape engineering of metal-organic frameworks', *Polyhedron*, 145, pp. 1–15. doi: 10.1016/j.poly.2018.01.004

111. Purewal, J. J. et al. (2012) 'Increased volumetric hydrogen uptake of MOF-5 by powder densification', *International Journal of Hydrogen Energy*, 37(3), pp. 2723–2727. doi: 10.1016/j.ijhydene.2011.03.002

112. Yuan, S. et al. (2017) 'PCN-250 under pressure: Sequential phase transformation and the implications for MOF densification', *Joule*, 1(4), pp. 806–815. doi: 10.1016/j.joule.2017.09.001

113. Bambalaza, S. E. et al. (2018) 'Compaction of a zirconium metal-organic framework (UiO-66) for high density hydrogen storage applications', *Journal of Materials Chemistry A*, 6(46), pp. 23569–23577. doi: 10.1039/C8TA09227C

114. Wang, T. C. et al. (2021) 'Surviving under pressure: The role of solvent, crystal size, and morphology during pelletization of metal-organic frameworks', *ACS Applied Materials and Interfaces*, 13(44), pp. 52106–52112. doi: 10.1021/acsami.1c09619

115. Yan, Y. et al. (2018) 'High volumetric hydrogen adsorption in a porous anthracene-decorated metal-organic framework', *Inorganic Chemistry*, 57(19), pp. 12050–12055. doi: 10.1021/acs.inorgchem.8b01607

116. Tian, T. et al. (2015) 'Mechanically and chemically robust ZIF-8 monoliths with high volumetric adsorption capacity', *Journal of Materials Chemistry A*, 3(6), pp. 2999–3005. doi: 10.1039/c4ta05116e

117. Connolly, B. M. et al. (2020) 'Shaping the future of fuel: Monolithic metal–organic frameworks for high-density gas storage', *Journal of the American Chemical Society*, 142(19), pp. 8541–8549. doi: 10.1021/jacs.0c00270

118. Madden, D. G. et al. (2022) 'Densified HKUST-1 monoliths as a route to high volumetric and gravimetric hydrogen storage capacity', *Journal of the American Chemical Society*, 144(30), pp. 13729–13739. doi: 10.1021/jacs.2c04608

119. Millange, F., Serre, C. & Férey, G. (2002) 'Synthesis, structure determination and properties of MIL-53as and MIL-53ht: The first CrIII hybrid inorganic-organic microporous solids: CrIII(OH)·(O2C-C6H4-CO2)·(HO2C-C6H4-CO2H)x', *Chemical Communications*, 2(8), pp. 822–823. doi: 10.1039/b201381a

120. Wang, X. et al. (2004) 'Syntheses, structures, and photoluminescence of a novel class of d 10 metal complexes constructed from pyridine-3,4-dicarboxylic acid with different coordination architectures', *Inorganic Chemistry*, 43(6), pp. 1850–1856. doi: 10.1021/ic035151s

121. Lee, S. J. et al. (2015) 'Uterine artery embolization for symptomatic fibroids in postmenopausal women', *Clinical Imaging*, 40(1), pp. 106–109. doi: 10.1016/j.clinimag.2015.08.010

122. Ibarra, I. A. et al. (2012) 'Near-critical water, a cleaner solvent for the synthesis of a metal-organic framework', *Green Chemistry*, 14(1), pp. 117–122. doi: 10.1039/c1gc15726d

123. Reinsch, H. et al. (2016) 'A facile "green" route for scalable batch production and continuous synthesis of zirconium MOFs', *European Journal of Inorganic Chemistry*, 2016(27), pp. 4490–4498. doi: 10.1002/ejic.201600295

124. Garcia Marquez, A. et al. (2013) 'Green scalable aerosol synthesis of porous metal-organic frameworks', *Chemical Communications*, 49(37), pp. 3848–3850. doi: 10.1039/c3cc39191d

125. Crawford, D. et al. (2015) 'Synthesis by extrusion: Continuous, large-scale preparation of MOFs using little or no solvent', *Chemical Science*, 6(3), pp. 1645–1649. doi: 10.1039/c4sc03217a

126. Didriksen, T., Spjelkavik, A. I. & Blom, R. (2017) 'Continuous synthesis of the metal-organic framework cpo-27-ni from aqueous solutions', *Journal of Flow Chemistry*, 7(1), pp. 13–17. doi: 10.1556/1846.2016.00040

127. Gao, H. L. et al. (2006) 'Synthesis and characterization of metal-organic frameworks based on 4-hydroxypyridine-2,6-dicarboxylic acid and pyridine-2,6-dicarboxylic acid ligands', *Inorganic Chemistry*, 45(15), pp. 5980–5988. doi: 10.1021/ic060550j

128. Głowniak, S. et al. (2021) 'Mechanochemistry: Toward green synthesis of metal-organic frameworks', *Materials Today*, 46(June), pp. 109–124. doi: 10.1016/j.mattod.2021.01.008

129. Batten, M. P. et al. (2015) 'Continuous flow production of metal-organic frameworks', *Current Opinion in Chemical Engineering*, 8, pp. 55–59. doi: 10.1016/j.coche.2015.02.001

130. Troyano, J. et al. (2020) 'Spray-drying synthesis of MOFs, COFs, and related omposites', *Accounts of Chemical Research*, 53(6), pp. 1206–1217. doi: 10.1021/acs.accounts.0c00133

131. Moosavi, S. M. et al. (2020) 'Understanding the diversity of the metal-organic framework ecosystem', *Nature Communications*, 11(1), p. 4068. doi: 10.1038/s41467-020-17755-8

132. Julien, P. A., Mottillo, C. & Friščić, T. (2017) 'Metal-organic frameworks meet scalable and sustainable synthesis', *Green Chemistry*, 19(12), pp. 2729–2747. doi: 10.1039/c7gc01078h

133. Shah, B. B., Kundu, T. & Zhao, D. (2019) 'Mechanical properties of shaped metal–organic frameworks', *Topics in Current Chemistry*, 377(5), p. 25. doi: 10.1007/s41061-019-0250-7

134. Cote, A. P. et al. (2005) 'Porous, crystalline, covalent organic frameworks', *Science. American Association for the Advancement of Science*, 310(5751), pp. 1166–1170.

135. Cote, A. P. et al. (2007) 'Reticular synthesis of microporous and mesoporous 2D covalent organic frameworks', *Journal of the American Chemical Society*, 129(43), pp. 12914–12915.

136. El-Kaderi, H. M. et al. (2007) 'Designed synthesis of 3D covalent organic frameworks', *Science. American Association for the Advancement of Science*, 316(5822), pp. 268–272.

137. Chae, H. K. et al. (2004) 'A route to high surface area, porosity and inclusion of large molecules in crystals', *Nature*, 427(6974), pp. 523–527.

138. Campbell, N. L. et al. (2009) 'Rapid microwave synthesis and purification of porous covalent organic frameworks', *Chemistry of Materials*, 21(2), pp. 204–206.

139. Uribe-Romo, F. J. et al. (2011) 'Crystalline covalent organic frameworks with hydrazone linkages', *Journal of the American Chemical Society*, 133(30), pp. 11478–11481.

140. Hunger, M. & Wang, W. (2006) 'Characterization of solid catalysts in the functioning state by nuclear magnetic resonance spectroscopy', *Advances in Catalysis*, 50, pp. 149–225.

141. Han, S. S. et al. (2008) 'Covalent organic frameworks as exceptional hydrogen storage materials', *Journal of the American Chemical Society*, 130(35), pp. 11580–11581.

142. Tylianakis, E., Klontzas, E. & Froudakis, G. E. (2011) 'Multi-scale theoretical investigation of hydrogen storage in covalent organic frameworks', *Nanoscale*, 3(3), pp. 856–869.

143. Guo, J.-H., Zhang, H. & Miyamoto, Y. (2013) 'New Li-doped fullerene-intercalated phthalocyanine covalent organic frameworks designed for hydrogen storage', *Physical Chemistry Chemical Physics*, 15(21), pp. 8199–8207.

144. Mendoza-Cortés, J. L., Han, S. S. & Goddard, W. A., III (2012) 'High H2 uptake in Li-, Na-, K-metalated covalent organic frameworks and metal organic frameworks at 298 K', *The Journal of Physical Chemistry A*, 116(6), pp. 1621–1631.

145. Furukawa, H. & Yaghi, O. M. (2009) 'Storage of hydrogen, methane, and carbon dioxide in highly porous covalent organic frameworks for clean energy applications', *Journal of the American Chemical Society*, 131(25), pp. 8875–8883.

146. Ding, S.-Y. & Wang, W. (2013) 'Covalent organic frameworks (COFs): From design to applications', Chemical Society Reviews. *Royal Society of Chemistry*, 42(2), pp. 548–568.

147. Peng, Y. et al. (2018) 'Hybridization of MOFs and COFs: A new strategy for construction of MOF@ COF core--shell hybrid materials', *Advanced Materials*, 30(3), p. 1705454.

148. Cai, M. et al. (2019) 'One-step construction of hydrophobic MOFs@ COFs core-shell composites for heterogeneous selective catalysis', *Advanced Science*, 6(8), p. 1802365.

149. Garzón-Tovar, L. et al. (2019) 'A MOF@ COF composite with enhanced uptake through interfacial pore generation', *Angewandte Chemie*, 131(28), pp. 9612–9616.

150. Feng, L. et al. (2020) 'Modular total synthesis in reticular chemistry', *Journal of the American Chemical Society*, 142(6), pp. 3069–3076.

151. Fan, H. et al. (2021) 'MOF-in-COF molecular sieving membrane for selective hydrogen separation', *Nature Communications*, 12(1), pp. 1–10.

152. Dawson, R., Cooper, A. I. & Adams, D. J. (2012) 'Nanoporous organic polymer networks', *Progress in Polymer Science*, 37(4), pp. 530–563.

153. Lee, J.-Y. et al. (2006) 'Hydrogen adsorption in microporous hypercrosslinked polymers', *Chemical Communications*, 25, pp. 2670–2672.

154. Ahn, J.-H. et al. (2006) 'Rapid generation and control of microporosity, bimodal pore size distribution, and surface area in Davankov-type hyper-cross-linked resins', *Macromolecules*, 39(2), pp. 627–632.

155. Tian, M. et al. (2019) 'Nanoporous polymer-based composites for enhanced hydrogen storage', *Adsorption*, 25(4), pp. 889–901.

156. Rose, M. et al. (2008) 'Element--organic frameworks with high permanent porosity', *Chemical communications*, 21, pp. 2462–2464.

157. Germain, J., Fréchet, J. M. J. & Svec, F. (2007) 'Hypercrosslinked polyanilines with nanoporous structure and high surface area: Potential adsorbents for hydrogen storage', *Journal of Materials Chemistry*, 17(47), pp. 4989–4997.

158. Rochat, S. et al. (2017) 'Hydrogen storage in polymer-based processable microporous composites', *Journal of Materials Chemistry A*, 5(35), pp. 18752–18761.

159. Ghanem, B. S. et al. (2007) 'A triptycene-based polymer of intrinsic microposity that displays enhanced surface area and hydrogen adsorption', *Chemical Communications*, 1, pp. 67–69.

160. Makhseed, S. & Samuel, J. (2008) 'Hydrogen adsorption in microporous organic framework polymer', *Chemical Communications*, 36, pp. 4342–4344.

161. Yuan, S. et al. (2009) 'Nanoporous polymers containing stereocontorted cores for hydrogen storage', *Macromolecules*, 42(5), pp. 1554–1559.

162. Hasell, T. et al. (2010) 'Palladium nanoparticle incorporation in conjugated microporous polymers by supercritical fluid processing', *Chemistry of Materials*, 22(2), pp. 557–564.

9 Analytical Methods, Modeling Approaches and Challenges of Nanomaterial-Based Hydrogen Storage

*Olugbenga Akande, Toheeb Jimoh,
Patrick U. Okoye, and Jude A. Okolie*

9.1 INTRODUCTION

Hydrogen is identified as a prospective alternative to conventional fossil fuels in addressing global energy demands, attributed to its abundance, high energy density, and capacity for generating emissions-free energy [1]. Nevertheless, the formidable challenge of safe and efficient storage impeded the widespread adoption of hydrogen. Nanomaterials have arisen as a solution to this challenge due to their distinctive characteristics. Using nanomaterials facilitates the safe storage and transport of hydrogen, as illustrated in Figure 9.1 [2]. Diverse categories of nanomaterials have been engineered to enable hydrogen storage, encompassing metal-organic frameworks (MOFs), carbon-based nanomaterials, complex hydrides, nanoporous materials,

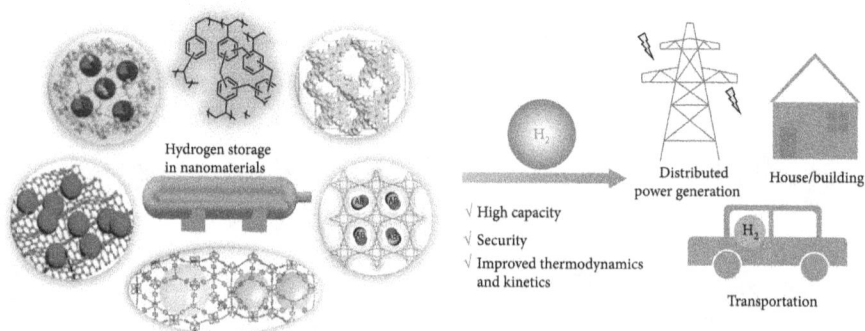

FIGURE 9.1   Sustainable applications of hydrogen stored in nanomaterials [2].

DOI: 10.1201/9781003371007-9

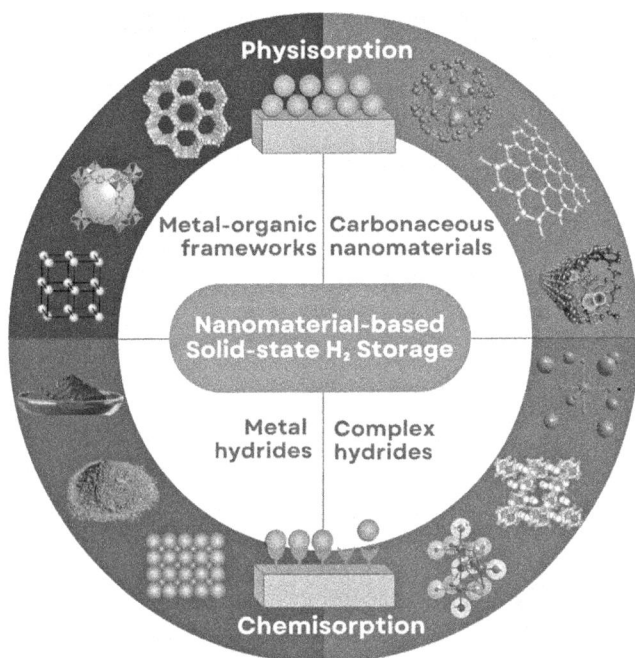

FIGURE 9.2   Types of nanomaterials used for the storage of hydrogen.

metal hydrides, and nanocomposites, as delineated in Figure 9.2. These nanomaterials exhibit distinctive benefits and limitations, leading to continuous research efforts to improve and innovate materials for hydrogen storage. Researchers have recently achieved notable advancements in nanomaterials tailored for hydrogen storage [3].

Strategies such as functionalization and surface modification, integrating hybrid materials, creating hierarchical structures, and implementing novel catalysts have been proposed to enhance the performance of nanomaterials in the context of hydrogen storage. Moreover, various characterization techniques, computational modeling, and simulation methodologies have been deployed to investigate nanomaterials for hydrogen storage [4]. These analytical approaches contribute to a comprehensive understanding of nanomaterials' intricate behaviors and properties, facilitating the ongoing refinement of materials for optimized hydrogen storage applications.

Nanomaterials present a promising avenue in addressing the complexities of hydrogen storage, with substantial advancements witnessed in the development of novel materials and enhancements to their performance. Nonetheless, deploying these materials requires thoughtful evaluation of their environmental and social implications [5]. Achieving a harmonious balance between technological progress and sustainable practices is imperative in shaping the trajectory of hydrogen storage technologies.

In conjunction with the practical techniques for hydrogen storage outlined by Abohamzeh et al. [6], the selection and integration of storage methods play a crucial role in establishing sustainable and efficient hydrogen-powered systems. While high-pressure storage, hydrogen liquefaction, chemical absorption, and physical adsorption represent diverse approaches, high-pressure tanks are the sole commercially available method for mobile systems [7].

Steel and aluminum, commonly used in traditional high-pressure hydrogen storage tanks, face limitations in strength [8]. Recent research has focused on developing composite tanks using materials like carbon fiber and epoxy, which exhibit high mechanical strength [6]. However, the high cost and maintenance requirements of these tanks have hindered their widespread adoption in hydrogen fuel cell-based technology [7]. Despite achieving a gravimetric density of 0.042 kg H_2/kg system at 700 bar pressure, the compressed storage method falls short of the gravimetric density target the US Department of Energy(DOE) set in 2020 [8].

Liquefied hydrogen, with a higher storage density than high-pressure gas storage, requires cooling systems and consumes more energy, leading to increased costs [7]. Double-walled cylinders with effective thermal insulation systems are deemed necessary to address hydrogen losses due to vaporization [6]. Chemisorption, involving metal hydrides maintaining hydrogen in a solid state, offers high storage density and stability with less space requirement. However, it suffers from irreversible hydrogen adsorption and desorption, necessitating high temperatures for desorption [6].

Physisorption, a more economical method, involves weak adsorption of hydrogen molecules at a material's surface through London dispersive forces. This method is efficient at normal temperatures and low pressures [7]. Physical adsorption, a widely researched method using porous materials, provides advantages such as high storage capacity, reversibility, fast charging and discharging, and low energy requirements [9]. Compared to cryogenic liquid storage, physical adsorption reduces energy requirements during charging and discharging, minimizing hydrogen boil-off losses [10].

Porous materials like zeolites, MOFs, and carbon materials, including fullerenes, nanotubes, and graphene, exhibit unique properties due to their high surface area, porosity, and uniform pore structure [11]. Increasing the porosity and homogeneity of these materials can enhance their hydrogen storage capacity. Therefore, physical adsorption using porous materials is crucial in overcoming current hydrogen storage challenges and achieving DOE's target value [12].

Integrating nanomaterials into this comprehensive landscape of hydrogen storage methods offers a potential solution due to their high surface area, enabling the storage of a large amount of hydrogen in a compact volume [13, 14]. The compatibility of nanomaterials with physisorption and chemisorption provides a versatile approach, presenting an avenue for efficient hydrogen storage in diverse applications.

Table 9.1 summarizes the advantages and limitations of different nanomaterials for hydrogen storage.

The present chapter discusses the analytical methods for evaluating hydrogen storage, different modeling approaches, and general issues associated with nanomaterial-based hydrogen storage.

TABLE 9.1

Summary of Merits and Limitations of Various Nanomaterials for Hydrogen Storage

Hydrogen Storage Materials	Advantages	Limitations	References
Metal Hydrides	1. High Hydrogen Storage Capacity 2. Safe and Stable 3. Reversible	1. Slow Kinetics 2. High Operating Temperatures 3. Limited Hydrogen Uptake Rates	Fan et al. [13]
Carbon Nanomaterials	1. High Surface Area 2. Safe and Stable 3. Reversible	1. Low Hydrogen Uptake Rates 2. Difficulty in Large-Scale Production 3. Challenges in Hydrogen Adsorption	Singh et al. [15]
Metal-Organic Frameworks (MOFs)	1. High Hydrogen Storage Capacity 2. Tunable Properties 3. Reversible	1. Limited Stability 2. Slow Kinetics 3. Limited Scalability	Li et al. [1]
Nanoporous Materials	1. High Surface Area 2. Tunable Properties 3. Safe and Stable	1. Limited Scalability 2. Challenges in Hydrogen Adsorption	Sepehri et al. [9]

9.2   CHALLENGES ASSOCIATED WITH NANOMATERIALS FOR HYDROGEN STORAGE

The research and development of nanomaterials for hydrogen storage have constituted an active and sustained focus for several decades. Notable among these nanomaterials are metal hydrides, carbon nanomaterials, MOFs, and nanoporous materials, distinguished by their elevated surface area and customizable properties. Despite their inherent promise, translating these materials into practical applications within onboard hydrogen storage and fuel cell systems remains a formidable undertaking, marked by numerous technical and economic obstacles. This section delves into the challenges associated with utilizing nanomaterials for hydrogen storage and provides an overview of the ongoing research efforts to overcome these challenges.

9.2.1   TECHNICAL CHALLENGES

The challenges associated with nanomaterials for hydrogen storage are multifaceted, with one primary hurdle being the slow kinetics of hydrogen uptake and release [12]. Despite their high surface area, the diffusion of hydrogen molecules into and out of nanomaterials can be hindered by factors such as pore size and surface chemistry. The imperative task involves enhancing the kinetics of hydrogen uptake and release, a crucial step for practically implementing these materials in both onboard hydrogen storage and fuel cell systems.

Another substantial challenge arises from nanomaterials' stability issues under diverse operating conditions [16]. Materials like metal hydrides and MOFs may undergo structural changes or degradation at elevated temperatures and pressures. Improving the stability of nanomaterials becomes paramount to ensure their long-term performance and reliability in hydrogen storage applications.

The scalability of nanomaterials poses a considerable challenge in commercial production, primarily due to elevated production costs and difficulties in maintaining consistent quality [17]. While laboratory-scale synthesis is attainable, achieving commercial viability remains a formidable task. Overcoming these production barriers is essential for successfully commercializing nanomaterials for hydrogen storage applications.

Safety considerations add another layer of complexity, given hydrogen's high reactivity and flammability as a fuel [16]. Ensuring the safe storage and transportation of hydrogen is imperative, requiring the design of nanomaterials for hydrogen storage with safety at the forefront. The prevention of hydrogen release and the minimization of associated risks are critical objectives in developing and utilizing nanomaterials for hydrogen storage systems. Addressing these challenges will collectively contribute to advancing and implementing nanomaterials in hydrogen storage technology.

9.2.2 ECONOMIC CHALLENGES

The commercialization of nanomaterials for hydrogen storage faces formidable economic barriers, primarily from the high costs associated with their production and utilization [15]. Factors such as the high cost of raw materials, substantial energy consumption, and the need for specialized equipment contribute to the overall financial burden. Addressing these cost-related challenges is imperative, necessitating the development of cost-effective production methods to facilitate the widespread adoption of nanomaterials for hydrogen storage.

Adding to the economic considerations is the concern surrounding the durability of nanomaterials for hydrogen storage, which introduces another layer of financial challenge [18]. Over time, the performance of these materials can degrade, leading to diminished efficiency and increased costs associated with replacement and maintenance. Developing durable nanomaterials capable of withstanding the harsh conditions related to hydrogen storage and utilization is essential for mitigating overall costs and ensuring sustained efficiency.

Recycling further underscores the economic challenges tied to nanomaterials for hydrogen storage that demand resolution. Many of these materials harbor valuable metals and other resources that, if recovered and reused, could reduce costs and minimize their environmental impact [12]. Thus, developing efficient recycling methods is a critical initiative to alleviate production costs and enhance the sustainability of nanomaterials for hydrogen storage. By addressing these economic challenges comprehensively, the path to the commercial viability of nanomaterials in hydrogen storage applications becomes clearer, fostering a more sustainable and cost-effective future.

9.2.3 CURRENT STATE OF RESEARCH

Despite the challenges associated with nanomaterials for hydrogen storage, significant progress has been made in addressing these challenges. Researchers have developed several strategies for improving these materials' kinetics, stability, scalability, and safety. Some of the current research directions include the following:

Catalysts are being developed to improve the kinetics of hydrogen uptake and release in nanomaterials for hydrogen storage [19]. These catalysts can enhance the surface chemistry of the materials, promote hydrogen dissociation, and improve hydrogen diffusion. For example, researchers have developed catalysts based on nanoparticles of palladium, platinum, and other metals, which can promote hydrogen adsorption and desorption in metal hydrides and other materials [20, 21]. Additionally, surface modification of carbon nanomaterials with metal oxide nanoparticles has improved the kinetics of hydrogen adsorption and desorption [22].

Novel materials are being designed and synthesized to address the scalability and durability of nanomaterials for hydrogen storage [23]. For example, researchers have developed MOFs with tunable pore sizes and structures that can enhance hydrogen uptake and release kinetics [24]. Additionally, using nanoporous materials such as zeolites and mesoporous silica can improve the scalability and durability of hydrogen storage materials [3].

Safety measures are being developed to mitigate the risk of hydrogen release and minimize the risk of accidents [25]. These measures include pressure relief valves, flame arrestors, and other safety devices to prevent the release of hydrogen in the event of an accident. Additionally, researchers are exploring using additives and coatings to improve the safety of nanomaterials for hydrogen storage [26].

Advanced characterization techniques are employed to examine the properties of nanomaterials designed for hydrogen storage at the atomic and molecular scales [9]. These methods offer valuable insights into such materials' surface chemistry, structure, and reactivity, thereby contributing to the enhancement of material development. Commonly utilized techniques include X-ray diffraction, transmission electron microscopy (TEM), and nuclear magnetic resonance (NMR) spectroscopy, which facilitate the investigation of the structure and properties of nanomaterials tailored for hydrogen storage [9].

9.3 OVERVIEW OF ADVANCED CHARACTERIZATION TECHNIQUES

Integral to contemporary materials science, advanced characterization techniques provide intricate insights into materials' nanoscale structure, composition, and properties. This detailed information is essential for comprehending material behavior and facilitating the development of novel materials with enhanced properties. Within the realm of hydrogen storage, numerous commonly employed characterization techniques are instrumental in evaluating the performance of nanomaterials. This section delves into an elucidation of these techniques.

9.3.1 Scanning Electron Microscopy

Scanning electron microscopy (SEM) is a well-established methodology for achieving high-resolution imaging of material surfaces [27]. In SEM, a directed electron beam is precisely focused onto the sample's surface, inducing the emission of secondary electrons. Detecting these emitted electrons creates an image depicting the material's surface [28]. The distinguished attributes of SEM, namely its high resolution and three-dimensional imaging capacity, render it an invaluable tool for analyzing surface features that may influence the mechanical properties of a material [29].

9.3.2 Transmission Electron Microscopy

Transmission electron microscopy (TEM) is a technique designed for imaging materials at the atomic scale [30]. A focused electron beam transverses a thin sample in the TEM process, with the transmitted electrons collected on a detector. This methodology enables researchers to visualize the internal structure of materials with atomic precision. TEM possesses several advantages over alternative imaging techniques, boasting high resolution and the capacity to furnish information concerning the crystal structure and defects within materials [30]. Moreover, it facilitates the study of material behavior under diverse conditions, such as elevated temperature or pressure, and can be employed to ascertain material morphology, size, and surface quality in terms of homogeneity and defects [31].

Shet et al. [32] provided a comprehensive overview of the current state-of-the-art hydrogen storage within MOFs. Their discussion encompassed various characterization techniques commonly applied to assess MOFs' performance in hydrogen storage, including X-ray diffraction (XRD), scanning electron microscopy (SEM), TEM, and thermal gravimetric analysis (TGA). The review highlighted recent advancements in the design and synthesis of MOFs, specifically aimed at enhancing their hydrogen storage capabilities. In a study by Zhang et al. [33], porous MOFs were synthesized and evaluated for hydrogen storage properties using XRD, N2 adsorption/desorption isotherms, and TGA. The findings demonstrated commendable hydrogen storage capacities and thermal stability, positioning these MOFs as promising candidates for hydrogen storage applications.

Furthermore, Suh et al. [34] evaluated hydrogen adsorption properties in various MOFs using a high-pressure volumetric method. Employing techniques such as XRD, SEM, and Brunauer–Emmett–Teller (BET) surface area analysis, they characterized the structures and properties of the MOFs. The outcomes indicated substantial hydrogen storage capacities and thermal stability, indicating the potential of these MOFs for hydrogen storage applications.

9.3.3 X-ray Diffraction

X-ray diffraction (XRD) is a widely employed technique in materials analysis for scrutinizing the crystal structure [35]. This method furnishes insights into a material's crystal structure and phase, offering sensitivity to alterations in the crystal structure induced by factors such as temperature, pressure, or other environmental variables [36]. Moreover, XRD proves valuable for identifying material impurities or defects [35]. It facilitates the determination of crystal structure, crystal size, and lattice strain, enabling the assessment of material purity, crystallinity, and structural integrity. This technique is instrumental in elucidating the bonding type and ascertaining the material's hydrogen content [36].

In a study by Zhou et al. [37], a 3D porous graphene aerogel was synthesized using a facile and scalable method. The hydrogen storage properties of the graphene aerogel were systematically characterized utilizing a combination of techniques, including XRD, transmission electron microscopy (TEM), and Raman spectroscopy. The outcomes revealed a notable hydrogen storage capacity and stability in the graphene aerogel. In a separate investigation, Mohan et al. [38] synthesized several nanostructured carbon materials, encompassing graphene oxide, carbon nanotubes, and activated carbon. These materials' hydrogen storage properties were evaluated through various techniques, including X-ray photoelectron spectroscopy (XPS), XRD, and scanning electron microscopy (SEM). The findings demonstrated high hydrogen storage capacities and stability in the nanostructured carbon materials.

9.3.4 RAMAN SPECTROSCOPY

Raman spectroscopy is a potent methodology for scrutinizing the vibrational modes of molecules within a material [39]. This technique furnishes insights into a material's chemical composition bonding characteristics and facilitates the identification of impurities or defects while enabling real-time monitoring of chemical reactions [40]. The Raman spectroscopy process involves directing a laser onto a material, with the analysis of the scattered light providing details about the material's vibrational modes [39]. Raman spectroscopy is non-destructive and can be applied to diverse materials, encompassing solids, liquids, and gases [40].

9.3.5 ATOMIC FORCE MICROSCOPY

Atomic force microscopy (AFM) is a robust technique for achieving high-resolution imaging of material surfaces [41]. This methodology delivers detailed images of surface topography and mechanical properties, facilitating the study of material behavior under different conditions, such as varying temperature or humidity [42]. The AFM process involves scanning a sharp probe over the material's surface, with the deflection of the probe utilized to generate a character image [41]. AFM is a non-destructive technique adaptable to diverse environments, including air, vacuum, and liquid settings [42].

9.3.6 THERMAL ANALYSIS

Thermal analysis techniques, including thermogravimetric analysis (TGA) and differential scanning calorimetry (DSC), are widely utilized for assessing the hydrogen storage capacity of materials, alongside examining the kinetics of hydrogen storage and release [33, 43]. Moreover, these techniques play a crucial role in evaluating the thermal stability of materials and elucidating their reaction pathways [33]. In TGA, the weight loss of a material is calculated as a function of temperature or time, providing information about the thermal stability and decomposition of the material [43]. Meanwhile, differential scanning calorimetry quantifies the heat flow associated with thermal events, such as phase transitions, and facilitates the determination of a material's enthalpy, entropy, and activation energy [33].

The application of thermal analysis techniques proves advantageous for investigating the hydrogen storage properties of materials, given the influence of temperature and pressure on hydrogen storage dynamics [43]. Furthermore, these techniques are non-destructive and can be employed under various conditions, enhancing their value as indispensable tools in materials science research.

9.4 COMPUTATIONAL MODELING AND SIMULATION OF NANOMATERIALS FOR HYDROGEN STORAGE

Nanomaterials designated for hydrogen storage have emerged as a promising energy storage and conversion solution, emphasizing the paramount importance of their meticulous design and optimization to attain elevated storage capacity and efficiency. Computational modeling and simulation have evolved into indispensable tools, enabling the prediction and comprehension of nanomaterial properties and behaviors at the atomic and molecular levels. This section will delve into computational methods employed in nanomaterials research, encompassing density functional theory, molecular dynamics simulations, Monte Carlo simulations, and machine learning approaches. The subsequent discussion will revolve around their applications in designing and optimizing nanomaterials tailored for hydrogen storage.

9.4.1 DENSITY FUNCTIONAL THEORY (DFT)

Density functional theory (DFT) is a prevalent quantum mechanical method extensively employed for predicting materials' electronic structures and properties at both atomic and molecular levels. Anchored in the principle that the electronic energy of a system can be deduced from the electron density, DFT entails solving the Schrödinger equation concerning the electron density. Within the domain of nanomaterials research, DFT has found application in investigating hydrogen adsorption on diverse materials, such as carbon nanotubes, MOFs, and porous materials. By predicting the binding energy and adsorption sites of hydrogen molecules on these materials, DFT calculations yield valuable insights into the mechanisms governing hydrogen storage [34, 44, 45].

9.4.2 MOLECULAR DYNAMICS (MD) SIMULATIONS

Molecular dynamics (MD) simulations represent a prevalent approach for investigating material behavior at both the atomic and molecular levels over time. This technique encompasses the resolution of classical equations of motion for each atom in the system, considering the forces between atomic particles and their interaction with the environment. MD simulations find widespread application in scrutinizing the dynamics of hydrogen storage within various nanomaterials, including but not limited to carbon nanotubes, graphene, and metal hydrides. Additionally, MD simulations offer insights into the diffusion and mobility of hydrogen molecules within the material, alongside providing information about the material's thermodynamic properties [46, 47].

9.4.3   Monte Carlo (MC) Simulations

Monte Carlo simulations have found application in examining hydrogen molecule adsorption and diffusion on materials based on graphene, including graphene oxide and functionalized graphene. The outcomes of these simulations unveiled that hydrogen adsorption is notably influenced by the functional groups affixed to the graphene surface and defects. Moreover, the diffusion of hydrogen molecules was observed to be contingent on factors such as the size of the pores and the presence of defects on the graphene surface [48].

9.4.4   Grand Canonical Monte Carlo (GCMC)

Grand Canonical Monte Carlo (GCMC) serves as an extension of the Monte Carlo (MC) simulation method, enabling the simulation of systems with a variable number of particles [49]. Consequently, GCMC simulations prove valuable for investigating systems where the particle count is not fixed, such as scenarios involving gas-liquid interfaces or the adsorption of molecules onto surfaces. In GCMC simulations, the number of particles is treated as a random variable, and the probability of a specific configuration is computed using the Metropolis algorithm [50].

In a recent study by Karki and Chakraborty, GCMC simulations were employed to delve into hydrogen storage in carbon-based materials [51]. The researchers utilized GCMC simulations to model hydrogen adsorption on activated carbon. The outcomes of their study suggested that optimizing pore size and distribution could significantly enhance the hydrogen storage capacity of the material.

9.4.5   Kinetic Monte Carlo (KMC)

Kinetic Monte Carlo (KMC) is a simulation method for investigating the dynamics of systems comprising numerous particles [52]. This approach is grounded in the concept that discrete events dictate particle motion within the system, including diffusion, reaction, or adsorption. The probability of each event is computed, and the system progresses by randomly selecting and executing one of the events. Noteworthy is the utility of KMC simulations in scrutinizing systems characterized by intricate dynamics, exemplified in scenarios such as catalytic reactions or surface growth [26].

9.4.6   Machine Learning (ML) Approaches

Recently, the application of machine learning (ML) techniques to investigate nanomaterials for hydrogen storage has become increasingly prevalent [53]. Machine learning algorithms play a pivotal role in the development of predictive models for the adsorption and desorption properties of materials. These models are constructed using extensive datasets derived from computational simulations or experimental data [53]. Their utilization expedites identifying novel materials possessing the desired characteristics for efficient hydrogen storage. Furthermore, ML methodologies have been successfully employed to anticipate the properties of diverse nanomaterials, encompassing carbon nanotubes, graphene, and MOFs [54]. ML models are adept at forecasting various aspects of nanomaterials, such as adsorption capacity, selectivity, hydrogen storage stability, and their synthesis and processing conditions.

Despite the advancements facilitated by computational modeling and simulation, their dependency on the accuracy of underlying physical models and parameters employed in calculations is a notable constraint. Consequently, meticulous calibration and validation of these models are imperative to ensure the precision of predictions [24]. Notwithstanding these limitations, computational modeling and simulation are indispensable tools for designing and optimizing nanomaterials tailored for hydrogen storage. The virtual exploration of an extensive array of materials and operational conditions significantly accelerates the discovery of materials with high hydrogen storage capacities and favorable properties [55].

9.5 SUMMARY OF KEY POINTS

Nanomaterials represent a promising avenue for developing viable and efficient hydrogen storage solutions. Utilizing molecular dynamics simulations and density functional theory calculations has yielded valuable insights into nanomaterials' structural characteristics and properties intended for hydrogen storage. Researchers have explored catalysts to enhance the kinetics of hydrogen uptake and release in hydrogen storage materials. Recent strides in these domains have demonstrated encouraging outcomes in augmenting hydrogen storage materials' efficiency, stability, and selectivity. These improvements render such materials well-suited for diverse applications, including fuel cells, portable power sources, and hydrogen-powered vehicles.

9.5.1 Future Directions for Research in Nanomaterials for Hydrogen Storage

To augment the efficiency and feasibility of hydrogen storage, forthcoming research in nanomaterials should focus on creating materials with high hydrogen uptake and release kinetics, selectivity, and stability. This entails the design and synthesis of nanomaterials with unique properties and functionalities, along with improvements in computational and experimental methods for characterization and optimization. Achieving success requires collaborative efforts among academia, industry, and government institutions. While research in nanomaterials for hydrogen storage is still in its early stages, computational techniques such as molecular dynamics simulations and density functional theory calculations can contribute to their design and optimization. Additionally, there is a need for more efficient and cost-effective hydrogen storage catalysts.

The potential for hydrogen storage in MOFs is promising due to their high surface area and adjustable pore size. Future research should emphasize enhancing their hydrogen storage capacity and stability and exploring their use as catalysts for hydrogen production. Carbon-based materials, including activated carbon, carbon nanotubes, and graphene, also exhibit promise. They have demonstrated high hydrogen uptake and thermal stability, and ongoing efforts seek to improve their performance through novel synthesis methods and hybrid materials.

In pursuing advanced hydrogen storage solutions, future nanomaterial research should prioritize the development of novel materials with enhanced properties, optimizing nanomaterial properties for specific applications, integrating with other technologies, and the scale-up and commercialization of nanomaterials with improved performance and cost-effectiveness. By leveraging materials synthesis, characterization, and modeling advances, tailored materials can be designed for specific hydrogen storage applications, paving the way for a cleaner and more sustainable energy future.

9.5.2   Importance of Collaboration and Investment

The progress and deployment of nanomaterials for hydrogen storage require significant investment and cooperative endeavors. A collaborative approach among academia, industry, and government institutions, involving sharing knowledge, expertise, and resources, can facilitate the development of more effective and practical hydrogen storage solutions. Moreover, channeling investments into research and development, infrastructure, and policy incentives can expedite the incorporation of hydrogen as a clean and sustainable energy source. This integration, in return, has the potential to foster economic growth, enhance energy security, and promote environmental sustainability.

9.5.3   Implications for a Sustainable and Equitable Energy Future

Exploring nanomaterials for hydrogen storage represents a significant prospect for fostering a sustainable and equitable energy future. Using hydrogen as an energy source can substantially mitigate greenhouse gas emissions and reduce dependence on fossil fuels. Additionally, incorporating nanomaterials for hydrogen storage can elevate the efficiency and safety of hydrogen storage, making it a more feasible option for widespread adoption.

To fully unlock the capabilities of nanomaterials in hydrogen storage, it is essential to leverage computational methods and catalysts to assist in designing and optimizing these materials. Moreover, collaboration and investment will play a pivotal role in developing and commercializing nanomaterials for hydrogen storage. Ultimately, the responsible and fair utilization of hydrogen as an energy source is crucial for unlocking the potential of nanomaterials in hydrogen storage and playing a role in establishing a more sustainable global environment.

9.6   CONCLUSION

The advancement of nanomaterials for hydrogen storage holds the potential to revolutionize the energy industry by offering a clean, efficient, and sustainable energy source. However, these materials are associated with slow kinetics, stability issues, scalability, and safety concerns. Addressing these challenges necessitates a collaborative effort involving researchers, industry professionals, and policymakers. The current research is promising, marked by substantial progress in developing enhanced nanomaterials and addressing the technical and economic challenges linked to their implementation. Continuous research and development could position nanomaterials for hydrogen storage as a vital component in the transition to a low-carbon economy.

Depending on the material and the required information, various characterization methods can be employed to assess nanomaterials for hydrogen storage. The demand for sophisticated characterization methods is expected to rise as materials become more complex. Future research will concentrate on refining existing methods and devising new ones to gain more detailed insights into material behavior at the nanoscale. Integrating advanced characterization techniques with computational modeling enables a comprehensive understanding of materials and facilitates the design of new materials with tailored properties for specific applications. These advanced characterization techniques have played a transformative role in materials science, contributing significantly to developing materials with improved properties. By continuously advancing these techniques, researchers can uncover new material behaviors and push the boundaries of materials science.

Moreover, advanced characterization techniques contribute to developing sustainable and environmentally friendly materials. Evaluating the hydrogen storage capacity of materials is crucial for developing materials suitable for hydrogen fuel cells, offering clean energy solutions for transportation and various applications. Similarly, characterizing the properties of nanomaterials is essential for developing efficient and cost-effective catalysts for different chemical reactions with significant environmental benefits.

In conclusion, advanced characterization techniques have been instrumental in developing nanomaterials for hydrogen storage and other applications, offering valuable insights into the properties and behavior of materials at the nanoscale. Metal hydrides and carbon-based materials emerge as the most utilized nanomaterials, employing metal-hydrogen bonds for hydrogen storage in metal hydrides and physisorption in carbon-based materials. The performance of these materials in hydrogen storage is influenced by their composition, morphology, and surface area. Understanding their properties and mechanisms is pivotal for their effectiveness in hydrogen storage, with applications extending to fuel cells and hydrogen-powered vehicles.

REFERENCES

1. Li JR, Kuppler RJ, Zhou HC. Selective gas adsorption and separation in metal-organic frameworks. *Chem Soc Rev.* 2009;38(5):1477–504.
2. Zheng J, Wang CG, Zhou H, Ye E, Xu J, Li Z, et al. Current research trends and perspectives on solid-state nanomaterials in hydrogen storage. *Research* [Internet]. 2023 Jul 31;2021. Available from: https://doi.org/10.34133/2021/3750689
3. Boateng E, Chen A. Recent advances in nanomaterial-based solid-state hydrogen storage. *Mater Today Adv.* 2020;6:100022.
4. Nagar R, Srivastava S, Hudson SL, Amaya SL, Tanna A, Sharma M, et al. Recent developments in state-of-the-art hydrogen energy technologies – Review of hydrogen storage materials. *Sol Compass* [Internet]. 2023;5:100033. Available from: https://www.sciencedirect.com/science/article/pii/S2772940023000012

5. Zhao L, Xu F, Zhang C, Wang Z, Ju H, Gao X, et al. Enhanced hydrogen storage of alanates: Recent progress and future perspectives. *Prog Nat Sci Mater Int.* 2021;31(2):165–79.

6. Abohamzeh E, Salehi F, Sheikholeslami M, Abbassi R, Khan F. Review of hydrogen safety during storage, transmission, and applications processes. *J Loss Prev Process Ind.* 2021;72:104569.

7. Hwang HT, Varma A. Hydrogen storage for fuel cell vehicles. *Curr Opin Chem Eng.* 2014;5:42–8.

8. Kan HM, Zhang N, Wang XY, Sun H. Recent advances in hydrogen storage materials. *Adv Mater Res.* 2012;512:1438–41.

9. Sepehri S, Liu YY, Cao GZ. Nanostructured materials for hydrogen storage. *Adv Mater Res.* 2010;132:1–18.

10. Niaz S, Manzoor T, Pandith AH. Hydrogen storage: Materials, methods and perspectives. *Renew Sustain Energy Rev.* 2015;50:457–69.

11. Hu Y, Zhang L. Hydrogen storage in metalorganic frameworks. *Adv Mater.* 2010 May 25;22:E117–30.

12. Konni M, Dadhich AS, Mukkamala SB. Hydrogen uptake performance of nanocomposites derived from Metal-organic Framework (Cu-BTC) and metal decorated multi-walled carbon nanotubes (Ni@f-MWCNTs or Pd@f-MWCNTs). *Surf Interfaces* [Internet]. 2020;21:100672. Available from: https://www.sciencedirect.com/science/article/pii/S2468023020306647

13. Fan L, Lin S, Wang X, Yue L, Xu T, Jiang Z, et al. A series of metalorganic framework isomers based on pyridinedicarboxylate ligands: Diversified selective gas adsorption and the positional effect of methyl functionality. *Inorg Chem* [Internet]. 2021 Feb 15;60(4):2704–15. Available from: https://doi.org/10.1021/acs.inorgchem.0c03583

14. Zhang L, Allendorf MD, Balderas-Xicohténcatl R, Broom DP, Fanourgakis GS, Froudakis GE, et al. Fundamentals of hydrogen storage in nanoporous materials. *Prog Energy.* 2022;4(4):42013.

15. Singh R, Altaee A, Gautam S. Nanomaterials in the advancement of hydrogen energy storage. *Heliyon.* 2020 Jul;6(7):e04487.

16. Han Y, Ho WSW. Polymeric membranes for CO_2 separation and capture. *J Memb Sci* [Internet]. 2021;628:119244. Available from: https://www.sciencedirect.com/science/article/pii/S0376738821001940

17. Zeleňák V, Saldan I. Factors affecting hydrogen adsorption in metalorganic frameworks: A short review. *Nanomater* (Basel, Switzerland). 2021 Jun;11(7).

18. Srinivasan S, Demirocak DE, Kaushik A, Sharma M, Chaudhary GR, Hickman N, et al. Reversible hydrogen storage using nanocomposites. *Appl Sci.* 2020;10(13):4618.

19. Salman MS, Pratthana C, Lai Q, Wang T, Rambhujun N, Srivastava K, et al. Catalysis in solid hydrogen storage: Recent advances, challenges, and perspectives. *Energy Technol.* 2022;10(9):2200433.

20. An C, Liu G, Li L, Wang Y, Chen C, Wang Y, et al. In situ synthesized one-dimensional porous Ni@ C nanorods as catalysts for hydrogen storage properties of MgH_2. *Nanoscale.* 2014;6(6):3223–30.

21. Tian M, Rochat S, Polak-Kraśna K, Holyfield LT, Burrows AD, Bowen CR, et al. Nanoporous polymer-based composites for enhanced hydrogen storage. *Adsorption* [Internet]. 2019;25(4):889–901. Available from: https://doi.org/10.1007/s10450-019-00065-x

22. Salahu N, Haq B, Abdullah D, Shehri A, Al-Ahmed A, Mizanur M, et al. Hydrogen storage in depleted gas reservoirs: A comprehensive review. *Fuel* [Internet]. 2023;337(November 2022):127032. Available from: https://doi.org/10.1016/j.fuel.2022.127032

23. Hirose K. Hydrogen as a fuel for today and tomorrow: Expectations for advanced hydrogen storage materials/systems research. *Faraday Discuss.* 2011;151:11–8.
24. Jing Z, Yuan Q, Yu Y, Kong X, Tan KC, Wang J, et al. Developing ideal metal-organic hydrides for hydrogen storage: From theoretical prediction to rational fabrication. *ACS Mater Lett.* 2021;3(9):1417–25.
25. Rajaura RS, Srivastava S, Sharma PK, Mathur S, Shrivastava R, Sharma SS, et al. Structural and surface modification of carbon nanotubes for enhanced hydrogen storage density. *Nano-Struct Nano-Objects.* 2018;14:57–65.
26. Onishi N, Iguchi M, Yang X, Kanega R, Kawanami H, Xu Q, et al. Development of effective catalysts for hydrogen storage technology using formic acid. *Adv Energy Mater.* 2019;9(23):1801275.
27. Akhtar K, Khan SA, Khan SB, Asiri AM. Scanning electron microscopy: Principle and applications in nanomaterials characterization. In: Sharma, S. (eds) *Handbook of Materials Characterization*. Springer; 2018. pp. 113–45.
28. Mielańczyk Ł, Matysiak N, Klymenko O, Wojnicz R. Transmission electron microscopy of biological samples. *Transm Electron Microsc Appl.* 2015;193–236. Doi: 10.5772/60680
29. Zemberekci L, Demir G, Akaoglu C, Aldulaimi WA, Ozhan AB, Gulgun MA, et al. Polymer bridging induced by a single additive imparts easy-to-implement green machinability to yttria-stabilized zirconia. *ACS Appl Polym Mater.* 2021;3(11):5397–404.
30. Abou-Ras D, Kirchartz T, Rau U. *Advanced characterization techniques for thin film solar cells*. Vol. 2. Wiley Online Library; 2011.
31. Choudhary MK, Garg S, Kaur A, Kataria J, Sharma S. Green biomimetic silver nanoparticles as invigorated colorimetric probe for Hg2+ ions: A cleaner approach towards recognition of heavy metal ions in aqueous media. *Mater Chem Phys.* 2020;240:122164.
32. Shet SP, Priya SS, Sudhakar K, Tahir M. A review on current trends in potential use of metal-organic framework for hydrogen storage. *Int J Hydrogen Energy.* 2021;46(21):11782–803.
33. Zhang J, Li Z, Wu Y, Guo X, Ye J, Yuan B, et al. Recent advances on the thermal destabilization of Mg-based hydrogen storage materials. *RSC Adv.* 2019;9(1):408–28.
34. Suh MP, Park HJ, Prasad TK, Lim DW. Hydrogen storage in metal-organic frameworks. *Chem Rev.* 2012;112(2):782–835.
35. Holder CF, Schaak RE. *Tutorial on powder X-ray diffraction for characterizing nanoscale materials*. Vol. 13, ACS Nano. ACS Publications; 2019. p. 7359–65.
36. Bardella F, Montes Rodrigues A, Leal Neto RM. CrystalWalk: Crystal structures, step by step. *J Appl Crystallogr.* 2017;50(3):949–50.
37. Zhou L, Yang Z, Yang J, Wu Y, Wei D. Facile syntheses of 3-dimension graphene aerogel and nanowalls with high specific surface areas. *Chem Phys Lett.* 2017 Jun 1;677:7–12.
38. Mohan M, Sharma VK, Kumar EA, Gayathri V. Hydrogen storage in carbon materials—A review. *Energy Storage.* 2019;1(2):e35.
39. Smith E, Dent G. *Modern Raman spectroscopy: A practical approach*. John Wiley & Sons; 2019.
40. Ferrari AC, Basko DM. Raman spectroscopy as a versatile tool for studying the properties of graphene. *Nat Nanotechnol.* 2013;8(4):235–46.
41. Li J, Liu Y, Yuan Y, Huang B. Applications of atomic force microscopy in immunology. *Front Med.* 2021;15:43–52.
42. Sugimoto Y, Pou P, Abe M, Jelinek P, Pérez R, Morita S, et al. Chemical identification of individual surface atoms by atomic force microscopy. *Nature.* 2007;446(7131):64–7.
43. Kumar S, Sharma R, Murthy SS, Dutta P, He W, Wang J. Thermal analysis and optimization of stand-alone microgrids with metal hydride based hydrogen storage. *Sustain Energy Technol Assessments.* 2022;52:102043.

44. Szczęśniak B, Choma J, Jaroniec M. Gas adsorption properties of hybrid graphene-MOF materials. *J Colloid Interface Sci*. 2018;514:801–13.

45. Zheng Y, Wang E, Zhou J, Sun Z. A theoretical study of 0D Ti_2CO_2/2D g-C3N4 Schottky-junction for photocatalytic hydrogen evolution. *Appl Surf Sci*. 2023;616:156562.

46. Vijayaraghavan V, Dethan JFN, Garg A. Tensile loading characteristics of hydrogen stored carbon nanotubes in PEM fuel cell operating conditions using molecular dynamics simulation. *Mol Simul*. 2018;44(9):736–42.

47. Wu CD, Fang TH, Lo JY, Feng YL. Molecular dynamics simulations of hydrogen storage capacity of few-layer graphene. *J Mol Model* [Internet]. 2013;19(9):3813–9. Available from: https://doi.org/10.1007/s00894-013-1918-5

48. Jain V, Kandasubramanian B. Functionalized graphene materials for hydrogen storage. *J Mater Sci* [Internet]. 2020;55(5):1865–903. Available from: https://doi.org/10.1007/s10853-019-04150-y

49. Allen MP, Tildesley DJ. *Computer simulation of liquids*. Oxford University Press; 2017.

50. Frenkel D, Smit B. *Understanding molecular simulation: From algorithms to applications*. Academic Press San Diego; 2002.

51. Karki S, Chakraborty SN. Hydrogen adsorption in nanotube and cylindrical pore: A grand canonical Monte Carlo simulation study. *Int J Hydrogen Energy*. 2023;48(7):2731–41.

52. Gillespie DT. Stochastic simulation of chemical kinetics. *Annu Rev Phys Chem*. 2007;58:35–55.

53. Salehi K, Rahmani M, Atashrouz S. Machine learning assisted predictions for hydrogen storage in metal-organic frameworks. *Int J Hydrogen Energy* [Internet]. 2023; Available from: https://www.sciencedirect.com/science/article/pii/S0360319923022139

54. Rahnama A, Zepon G, Sridhar S. Machine learning based prediction of metal hydrides for hydrogen storage, part I: Prediction of hydrogen weight percent. *Int J Hydrogen Energy*. 2019;44(14):7337–44.

55. Wang B, Xie LH, Wang X, Liu XM, Li J, Li JR. Applications of metal-organic frameworks for green energy and environment: New advances in adsorptive gas separation, storage and removal. *Green Energy Environ*. 2018;3(3):191–228.

Index

For Product Safety Concerns and Information please contact our EU
representative GPSR@taylorandfrancis.com
Taylor & Francis Verlag GmbH, Kaufingerstraße 24, 80331 München, Germany

www.ingramcontent.com/pod-product-compliance
Lightning Source LLC
Chambersburg PA
CBHW070717220326
41598CB00024BA/3201